GEOLOGY:

ROCKS, MINERALS and LANDFORMS

Jack Fleming

Contents

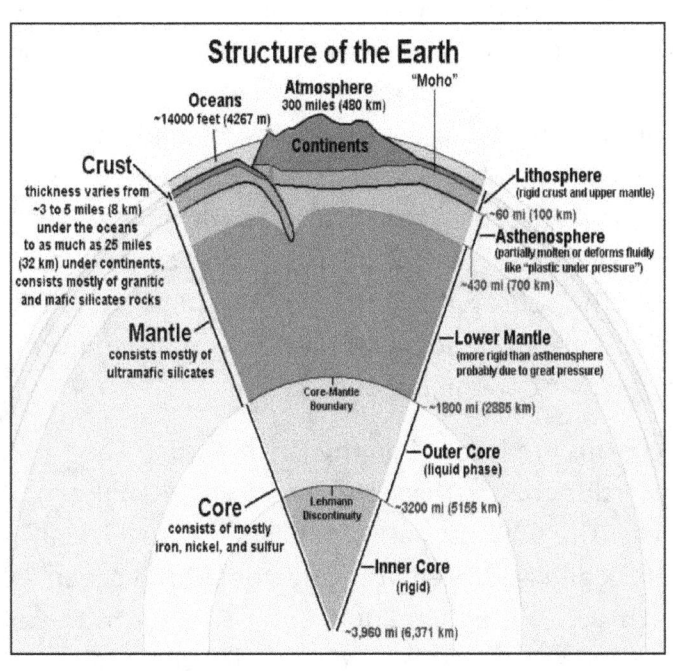

Structure of the Earth

Oceans
~14000 feet (4267 m)

Atmosphere
300 miles (480 km)

"Moho"

Continents

Crust
thickness varies from
~3 to 5 miles (8 km)
under the oceans
to as much as 25 miles
(32 km) under continents,
consists mostly of granitic
and mafic silicates rocks

Lithosphere
(rigid crust and upper mantle)

~60 mi (100 km)

Asthenosphere
(partially molten or deforms fluidly
like "plastic under pressure")

~430 mi (700 km)

Mantle
consists mostly of
ultramafic silicates

Lower Mantle
(more rigid than asthenosphere
probably due to great pressure)

Core-Mantle
Boundary

~1800 mi (2885 km)

Outer Core
(liquid phase)

Lehmann
Discontinuity

~3200 mi (5155 km)

Core
consists of mostly
iron, nickel, and sulfur

Inner Core
(rigid)

~3,960 mi (6,371 km)

CHAPTER 1   GEOLOGY

The term Geology is the study of the earth and comes from the Greek word 'geo' which means land or earth. The term 'logos' comes from the Greek word meaning story. Geology is the study of the solid features of our planet and the process of change over time. This includes the Earth's composition, processes, structure, and its history. Geology is the study of materials that make up the Earth, how changes occur, and the history of why the changes had occurred. Geologist determines the unique landscape. They formulate why hills and valleys are where they are and why some areas don't contain such features. They observe where rivers take a turn and why they do at that location. Geology is the study of all landforms from the frozen arctic to the dry deserts.

All landforms are located on the Earth's crust. The thickest part of the Earth's crust varies between 35 and 40 miles in thickness. The thinnest part of the Earth's crust is only 3 to 4 miles thick and is located at the bottom of the ocean. Below the Earth's crust is a thick liquid outer core forms the mantle. It consists primarily of molten iron and nickel. In the center of the mantle is a solid iron and nickel core. The mantle is approximately 1,800 miles think. The outer core is 1,380 miles thick and is fluid in nature. The inner core is solid and 780 miles in diameter. The Earth's diameter is 7,926 miles. Only 12 miles separates the Earth's highest point to its lowest point. The highest point is Mt. Everest which is 29,208 feet. The lowest point is the Mariana Trench in the south-west Pacific. It is 35,800 feet. Water covers 75% of the Earth's surface.

The Earth is a system of land, air, sea, animals, and plants. All are in constant change. Geology also involves other areas of science, such as geochemistry, physics, biology, paleontology, meteorology, crystallography, marine geology, stratigraphy (the study of sedimentary rock sequences in time periods), and geomorphology.

Earth's Crust - The outer solid crust surrounds the mantle. The crust makes up 0.6 percent of the Earth's volume and 0.4 percent of the Earth's mass. Beneath the ocean, the crust is chiefly composed of basalt. There is a variation of thickness between 3 and 6.8 miles (5 to 11 kilometers). The Earth's crust beneath the continents, the thickness, composed chiefly of granite, is between 12 to 40 miles (19 to 64 kilometers) thick. The lithosphere is the Earth's upper layer. It includes the continental, oceanic, and some of the cooler solid upper mantle.

Earth's Mantle – The mantle comprises 84 percent of the Earth's volume and 67 percent of the Earth's mass. It is 1,802 miles (2,900 kilometers) thick and contains silica, iron, magnesium, and metal-rich minerals. The uppermost portion of the Earth's core is separated by the Mohorovicic discontinuity. The Earth's mantle is separated from the outer core by the Gutenberg discontinuity.

The asthenosphere is the hot plastic part of the upper mantle and lower crust. T separates the brittle crust-mantle lithosphere above from the mesosphere below. This area is believed to be responsible for the continental movement of the crustal plates. This area is 186 miles (300 kilometers) thick. The mesosphere is below the asthenosphere and is more solid. It includes part of the upper and all of the lower mantle.

Earth's Core – Both the outer and inner core make up 15 percent of the Earth's by volume and 32 percent by mass. The inner core is 780 miles (1,287 kilometers) thick. The outer core is 1,380 miles (2,253 kilometers) thick. The inner core is located at the center of the Earth and is a solid mass. It is composed primarily of iron with a small amount of nickel. At sea level, the pressure is 1 atmosphere. At the core, the pressure is 3 million atmospheres. The temperature is between 7,200 and 9,000 degrees F. (4,000 to 5,000 C). That is almost as hot as the surface of the sun.

The composition of the Earth's Crust

Name	Mass in Percent
Oxygen	45.0
Silicon	27.0
Aluminum	8.0
Iron	5.8
Calcium	5.1
Magnesium	2.8
Sodium	2.3
Potassium	1.7
Hydrogen	1.5
Titanium	0.86
Phosphorus	0.1
Manganese	0.1
Florine	0.046
Strontium	0.045
Barium	0.038
Sulfur	0.030
Chlorine	0.019
Vanadium	0.017
Zirconium	0.014

Scientific Discoveries in Earth Science

B.C.   C.1000   Industrial use of iron in Mesopotamia and Egypt

c. 900   Natural Gas from wells used in China.

c. 390   Plato proposed that another continent existed across from   Europe.

c. 330   Pytheas suggested that tides were caused by the moon.

A.D.   c. 20   Stabo's Geography collects all known geographical information.

132   Zhang Heng invented the world's first crude seismograph

1086   Shen Kua's Dream Pool Essay described the principles of erosion, uplift and a movement of sedimentation.

1298   Marco Polo described coal and asbestos in Europe

1517   Girolamo Fracastoro described the remains of ancient organisms (fossils)

1546   Georgius Agricola coined the word 'fossil'.

1565   Konrad von Gesner produced the first drawings of fossils.

1743   Christopher Packe produced the first geologic map.

1774   The classification of minerals was introduced by Abraham Werner.

1780   The first fossil found to be that of a dinosaur was found in England.

1811   The first known fossil of an ichthyosaur was discovered in England by Mary Anning, 12 years old.

1821   The first known fossil of a plesiosaur was discovered by Mary Anning,   Then 21 years old.

1822   F. Mohs developed the scale for hardness of minerals.

1828   Paul Erman produces the first measurement of the Earth's magnetic field.

1837   The idea of 'ice ages' was produced by Louis Agassiz.

1842   The term 'dinosaur' is introduced by Richard Owen.

1850   Matthew Fontaine Maury draws the chart of the Atlantic Ocean.

1859   Edwin Drake drills the world's first oil well.

1860   Hermon von Meyer finds the earliest fossil of a bird.

1906   R. Oldham demonstrates earthquake waves are used to determine the Earth's layers.

1907   Rocks are dated using uranium by Bertram Bolrwood.

1909   Andrija Mohorovicic discovered Moho discontinuity.

1912   The continental drift theory is proposed by Alfred Wegener.

1914   Beno Cutenberg discovers the Gutenberg discontinuity.

1925   A German expedition discovers the Mid-Atlantic Ridge.

1935   The earthquake scale is developed by Charles Richter.

1958   Using satellite data, James Van Allen discovers the radiation belts.

1960   The United States first uses geothermal power. Harry Hess proses sea-floor spreading.

1968   The Glomar Challenger goes into service to study ocean depths.

1977   John Corliss and Robert Ballard discovered deep-sea vents near the Galapagos Islands.

1980   The idea that a large space object struck the Earth and caused the extinction of dinosaurs was first proposed by Luis Alvarez and others.

1985   The first deep-sea vents in the Atlantic Ocean is discovered by Peter Rona.

1987   Wade Miller finds the oldest known embryo from a fossil dinosaur egg.

1990   The oldest portion of the Pacific plate was discovered.

1991   Researchers found the first earthquake fault rupturing the surface in the northern North American region.

1992   The crater found in the Yucatan Peninsula related to the extinction of dinosaurs is discovered by scientist. Climate changes are recorded by scientist after they drilled into the ice in Greenland.

Earth Facts:

Size	Miles	Kilometers
Mean radius	3,958.7	6,370.95
Polar radius	3,950	6,332
Equatorial radius	3,964	6,378
Polar circumference	24,857	39,995
Equatorial circumference	24,900	40,064

Surface Area	Square Miles	Square Kilometers
Surface area	196,950,711	510,0501 x 10^6
Land area	57.5 million sq. mi.	149 x 10^6
Water area	139.4 million sq. mi	175,500 x 10^6

Escape Velocities	Miles per Second	Km per Second
From the surface	6.96 per second	11.2 per second
Maintain Orbit	66,672 per hour	107,280 per hour

Geological maps containing geological map symbols are usually created by geologist assembling information from direct observation, aerial photographs, and satellite data. Surface and subsurface features are observed to find fault lines and other surface features. Geologists take into account:

Dip or angle of the structural surface which is measured perpendicular to strike.

A strike is the compass direction of the geological feature.

Attitude is the flow or direction of the rocks.

SOLID MOVEMENT

The continents on Earth move both vertically and horizontally. Heat in the mantle creates a convection current that forms a type of blister which moves the mantle upward, then toward the sides. The melting of glaciers and erosion of rocks causes the crust to change position.

Geologist on igneous rock

IGNEOUS MOVEMENT

Heat is produced in the crust and mantle by the friction between moving blocks and radioactivity. This heat keeps the mantle in a fluid state as it melts some crustal rock. Some of this melting will cool and solidify in fractures and some will burst as volcanic activity. Below the surface, this hot material is called magma. When it flows as a volcano, it is referred to as lava. When the lava cools, it is referred to as igneous rock. There are two types of igneous rocks, intrusive and extrusive. When lava reaches the surface, it cools. This is extrusive igneous rock. When the lava cools quickly, the minerals form fine crystals. This type of rock, with the fine-grained minerals, is called basalt. In the United States, basalt flows are commonly found in the western part of the country. The intrusive igneous rocks cool slowly and much deeper into the Earth. Their coarse-grained mineral crystals are about a half inch in texture. Granite and diorite are examples of intrusive igneous rock.

Sometimes porphyry forms when the two stages of cooling occur. The large crystals are called phenocrysts and are found embedded in the finely grained mass. There are three identifiable characteristics of igneous rocks, the mineral content, the color, and texture. The texture shows the cooling time and depth of the igneous rock when it was formed. The kind of igneous rock is determined by the mineral content. The intrusive igneous rock is coarsely grained indicating that it cooled over a long period of time. They often reach the surface due to forces that uplift the material. The rocks are often weathered by wind and water. The color helps identify the magma from which the igneous rock was formed. The lighter colored igneous rocks such as granite, are usually formed in silicic magma. These specimens are rich in feldspar and quartz. The dark green or black basalts are formed from mafic magmas which are high in magnesium and iron.

GRADATION
Where the land is lowered by the destruction of rock, the rock waste is removed in areas called rock waste. This process is referred to a degradation. Where the rock is deposited is referred to as aggradation. This is where the land is raised. Gradation is primarily caused by erosion and deposition, weathering, and mass wasting.

WEATHERING
Rock disintegration is often a result on the surface due to weather conditions. Moisture, atmospheric gases, and chemicals attack the rock. Acids from decaying matter have an impact on rock. Water freezing in openings of the rock expands and breaks the rock apart. Growing plant life often has roots that find its way into the rock. The expansion of the plant

growth also contributes to the breaking down of rocks. Temperature changes cause the rock to expand or retract. This also breaks the rock formations. Weathering of rock on a hill falls into a ravine and that rock debris is called talus.

Freezing and thawing – Water freezes to form crystalline ice. This ice expands to a size larger than the water in its liquid form. The expanding ice exerts a force close to 2,000 pounds per square inch. This expansion can break rock apart. This ice-prying is the destructive force in the Arctic and mountain areas, especially where thawing and freezing occur frequently. This even produces most of the rock litter in that area. This frost heaving is often witnessed in the pavement, (pot-holes) and is a problem for communities. In arctic regions, the freezing of lake bottoms may cause soil-capped mounts called pingoes to rise as high as 100 feet.

In desert regions, rapid cooling of a rock surface may cause the rock to flake or release small grains of rock material to break off. The warmth of the daytime sun quickly drops in the evening or during a sudden rainstorm.

Organic activity can cause the change in the landscape. In warm, humid areas, plants grow rapidly and their roots take hold in rocks and soil. The rapid growth will cause small rock masses to break apart. These areas are also subject to animals that burrow in the ground. Their constant burrowing will have an effect on the landscape.

Chemical activity has an effect on the landscape. If the rock is exposed to dissolved gases and atmospheric water, minerals in the rock will decompose. This activity takes place in hot humid areas. The rock will either flake off or crumble. Limestone, gypsum, and marble are dissolved by the water containing carbon-dioxide. Sandstone and shale, consist of previously weathered rock waste will resist chemical weathering. Ferromagnesian produce hematite and limonite when they react to water. Feldspar reacts with water and carbon dioxide which tends to make silica and clay.

EROSION

Erosion is the wearing of rock activity which also transports and deposits the waste. The most erosive activity is water from streams. The power of moving water together with sediment acts as an abrasive agent wearing the rock. Underground water wears away rock forming cavities. Windblast the rock and carries its debris which may form dunes. Ocean currents and waves either erode the coastline or build up the beaches which may form beaches or sandbars. Glacier ice moving down mountain valleys, or over artic land, disintegrates the bedrock.

TIME in a geological sense

Originally it was thought that the shape of Earth's landscape was due to a cataclysmic event. However, scientists have determined that the events millions of years ago were no different than the geological events occurring today. Geologist James Hutton (1726-1797) formed the Hutton principle which is called uniformitarianism. By studying the geological changes now on Earth, a geologist would be able to determine Earth's geological past. Volcanoes and earthquakes can rapidly change the formation of an area, but most landforms are created over long periods of time.

The cone shape of a volcano is built by the lava erupting from its pipe. The pipe is the passage through the crust often where fractures intersect. Magma from a deep reservoir rises in the pipe and spews out on the surface. Sometimes the lava is blasted outward. In this case, a crater is formed in the cone by eruptions and the collapse of material around the vent when the lava in the pipe sinks. In either event, the lava is accumulated on the surface. Cones with a slope of 40 degrees or more is made of volcanic cinders that formed during mild explosions. Because of their loose construction and high permeability, they absorb rain and melting snow. Cones with a gentle slope are created by eruptions of basaltic lava at high temperatures, (2,200 degrees F.) with very little trapped gas to form an explosion. The lava in this condition is very fluid and may travel at 5 to 25 miles per hour while reaching 30 to 40 miles from the vent. A shield volcano cone is formed, sloping at 5 degrees at the summit. At the base, the slope would be around 12 degrees where the lava cools and builds up.

Meteorite crater

CHAPTER 3   PLATE TECTONICS – FOLDS AND FAULTS

Engineers depend on the geologist analyst before they plan to build tunnels, bridges, buildings, roads, wells, and other structures upon the land. Geologists draw geological maps from the information obtained from satellite data, aerial photographs, and direct observations in the field. The maps help geologist locate faults and direct field observations permit geologist to measure subsurface and surface features. Geologists observe attitude, strike, and dip. Attitude is the direction the rocks tend to move. A strike is the compass direction of the movement. A dip is an angle that the surface structure is compared to the ground surface. Surface expressions, or outcrops, are usually found on a topographical map. The valleys, hills, and stream direction are indicated on the map similar to a contour map. The outcrops indicate folds which are ground areas that are bent under extreme pressure. Rock and land formations follow certain patterns. A geologist can apply the principles of superposition, structural geology, and stratigraphy, and determine the formation, age, and changes below the Earth's surface.

Small particles of soil and rock fall from the suspended water of a stream, river, or ocean. The particles are deposited in layers. The law of superposition states that the older layers are below the bottom layers. Although the younger layers may be removed from time to time through erosion, new younger layers are then applied on top of the older layers.

Occasionally, rocks form a break in the layers. They are usually caused by pressure or erosion. When a series of rock layers are cut across, this land formation is referred to a fault. There are three forms, superposition, which is the younger layers on the top, stratigraphy, which is how layers form, and structural geology, which is how the layers change by folding. With this information, geologists determine what is below the surface.

Here, tectonic plates meet

Experiment:

Fill a clear container with rocks, soil, sand, and gravel. Fill the container with water. Shake the container for a minute or so. Allow it to sit still. Observe the layers that have formed. Did the largest material settle to the bottom? Why? Examine the top layers closely. What did you observe?

In the area of young mountain systems, the crust reacts to force on it and within it. The forces acting on the area is compression, torsional, tensile, and shearing. Compression involved the squeezing force. Torsional force is the twisting of the land. Tensile is the stretching of the area. Shearing is the pulling in opposite directions of the land. These forces cause faulting as the land breaks and become dislocated. The forces also cause folding which is the distortion of the land. This usually occurs over long periods of time and in small occurrences. These events produce relief of the pressure imposed. Faulting and folding will often produce areas in which eroding occurs. Most areas that are erosional have been induced by the faulting and folding in the area. Faults and folds are usually concealed or eroded. They can be spotted at quarries, road cuts or where the bedrock is exposed.

Folds – Rock doesn't necessarily crumble yet can be distorted. When pressure is applied to rock over a long period of time, permanent bends occur. In stratified rock, the layers slide over each other. Sometimes minerals recrystallize and align the layers parallel to the fold axis. Some folds may be hundreds of miles long while other folds may be extremely small in size.

Fold types:

Monoclines are local steep dips in the rock layer. They are almost horizontal sedimentary strata.

Anticlines are up-folds which form a type of arch. They are joined with synclines in groups of straight or slowly curing parallel axes.

Warps are minor distortions caused by the bending of land. This occurs in layered rocks that are raised or lowered.

Synclines are the trough area or the bottom of a "V" shaped down-fold. This occurs with an anticline.

Recumbent folds are axial plane areas and limbs that are almost horizontal which appear to be lying on its side.

Overturned Folds are folds having one limb rotated more than 90 degrees from the original horizontal. This type of fold also is also identified as an axial plane inclined where both limbs slope in the same direction.

Homoclines are masses of strata that dip in the same direction. This is not necessarily caused by folding. This may be a result of broad warping has been followed by faulting.

Plunging Folds are axes inclined to horizontal. A fold plunging at both ends is doubly plunging. These folds include one or more domes.

Folds induce relief by directing erosion. Causes of gradation begin to erode anticlines as they rise above sea level. Erosion may be a single water gap cut or the removal of several layers which would expose a full anticlinal ridge. Long periods of erosion develop ridges on upturned edges at the resistant strata and valleys in weak bands. Groups of large folds form mountains. Where differential erosion accounts for relief, anticlinal valleys and synclinal ridges are usually found. This forms an inverted topography.

Anticlinal Ridges are summits that are the crest of anticlines. They are usually rounded and broad in shape. A lower resistant layer caused by deep erosion may be exposed.

Anticlinal valleys are the following axis of an anticline. Their sides are usually steep.

Synclinal Ridges are summits at the synclinal trough. They are usually flat in shape and some have a lengthwise hollow. They have steep sides with a blunt end. They are found in areas left by deeper erosion of an adjoining anticline.

Synclinal Valleys are usually found following the axis of a syncline. The slopes are usually gentle.

Monoclinal Ridges and Valleys are formed by erosion of beds dipping in the same direction. The slopes are usually unequally steep.

Domes are up-folds with very little elongation along the axis. They form circles or ovals. The topography may be risen or inverted. The dome form is created from the vertical uplift where the old igneous rock is circled by younger, tilted sedimentary strata. As the top is reduced, a central basin may form. Streams may run down the sides toward the center as the flow joins with mainstreams and follow the valleys which cut through the basin's sides. On the outer slopes, the streams run toward the edges. Other streams flow in homoclinal valleys in the weaker belts that form a ring around the dome.

Faults are fractures with the displacement of adjoining crustal masses. When a fault movement occurs, it is followed by an earthquake. Most of the time this displacement is measured in inches. The 1964 earthquake in Alaska was 40 feet. The number of displacements over millions of years can product mountain ranges or large depressions.

Fault types:

Normal or gravity. The wall drops steeply in relation to the footwall.

Over-thrust fault shows many miles of displacement at a low angle.

A reverse thrust fault is shown with the hanging wall riding up the footwall. The footwall has dropped steeply.

Tear Strike-slip fault shows the movement is horizontal along the steeply dipping fault plane.

Faulting directly produces relief. The up-thrown block showing the exposed side is a fault scarp. It could be inches or miles high. This event occurs above and below sea level. A dormant fault may resume a parallel line creating a ramp joining the up-thrown and down-thrown blocks. Ramps have a limitation to recent fault scarps. Old scarps are high above sea level and the most eroded part is near the top. The younger part is located toward the bottom. Fault scarp formations are between the valleys and are triangular. The bare rock exposed on the scarp usually will have polished grooves because of abrasion as movement occurs along the fault plane. Scarps made by over-thrusts and thrust weather and erode waste to the stable slopes opposite to the dip of the fault. These scarps lack slick sides along the fault face.

Fault troughs are elongated depressions caused by faulting. One fault trough occurs along a single fault near two tilt blocks. A trough consists of a fault scarp with a gentle back-slope of a block. The second fault trough type is the graben. This type has two faults with the block being depressed and little tilting. Grabens may be many miles wide or many miles long.

The total picture of landforms

The Earth is always changing

CHAPTER 4   EARTHQUAKES

Earthquakes are the most dangerous events that destroy property and human life. Most of the time, the earthquakes occur in fault zones. This is where two plates come together, slip by each other, or split. The movement of the two plates causes great friction. When the two plates become locked, the forces push the plates in the direction of least resistance. This movement causes the earthquake. In areas with volcanoes, the heat contributes to the pressure which causes the tremors and localized quakes. How deep does the earthquake occur? Usually no more than 62 miles (100 Km) beneath the surface. The area where the quake occurs below the surface is called the focus. The area above the surface where the earthquake occurs is called the epicenter.

When the earthquake occurs, seismic waves travel in all directions from the focus. The fastest seismic waves are called the Primary waves or P-waves. These waves move back and forth. They are the fastest waves and can reach the far side of the Earth within twenty minutes. These waves travel through the Earth's molten core. The next series of seismic waves are called Secondary waves or S-waves. These waves also move side to side but travel through the solid portion of Earth. The next set of seismic waves are called the Love and Raleigh waves or L-waves. These waves move back and forth and up and down. These waves only occur at the Earth's surface. The L-waves are what causes the most damage during an earthquake.

Tsunamis

Underwater earthquakes or volcanoes often cause a tsunami which is a seismic sea wave. Most of the tsunamis occur in the Pacific Ocean. About 80 percent. The Atlantic Ocean experiences about 10 percent of the tsunamis and the remaining 10 percent occur in other oceanic regions. In the deep ocean, a tsunami is difficult to detect since the wave is only about one to two feet. The wavelength, crest to crest can range about 600 miles (965 Km). The primary damage of a tsunami occurs as the wave approaches the coast. The tsunami may be as high as 200 feet (61m). These tsunamis can impact the coast at 150 miles per hour (24 Km/h). The impact of this amount of seawater at this speed causes much destruction.

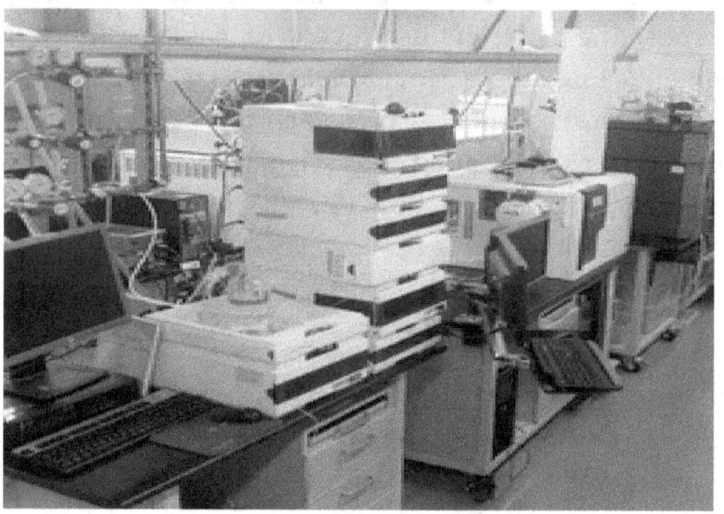

Earthquake Lab

In 1902, seismologist Giuseppe Mercalli developed the Mercalli earthquake intensity scale. It measured the Earth's destructiveness.

Level   Effects of destruction

1   Minor tremor only detected by a seismograph

2   Objects may swing. Only a few people may detect the quake

3   Sounds like a passing truck. It is usually detected by people indoors

4   Windows and dishes rattle. It is detected by a number of people

5   Items fall. Sleeping people are awakened. Doors move. Felt by everyone

6   Bushes and trees shake. Windows rattle or break. Felt by everyone. People will walk with difficulty

7   Unable to stand. Moderate to heavy damage to buildings. Walls crack.

8   Tree branches break. Chimneys and walls collapse. Unable to steer an automobile

9   Foundations crack. Extensive damage to buildings. Underground pipe break

10   Most masonry, frame structures, and foundations are destroyed. Railroad tracks bend. Water is thrown on banks of rivers.

11   Few building remain. Railroad tracks are destroyed. Bridges are destroyed. Underground pipes are destroyed

12   Nearly total destruction; large rock masses are displaced. Objects are thrown into the air.

The Richter scale

In 1935, the Richter scale was developed by Charles Richter. The Richter scale measures the earthquake's magnitude or intensity on a scale from 1 to 8.8. The logarithmic scale is measured where each successive whole number is increased 10 fold in power. Each number represents the maximum amplitude of the seismic wave at a distance of 100 miles (161 Km). The distance in time between the first and second wave is measured. An empirical factor is added. This takes into account that the wave grows weaker from the focus point.

Richter Number	Increase in Magnitude
1	1
2	10
3	100
4	1,000
5	10,000
6	100,000
7	1,000,000
8	10,000,000

Energy Comparison

Richter Number	Approximate Energy Returned (TNT equivalent)
1	170 grams of TNT
2	6 Kilograms of TNT
3	179 Kilograms of TNT
4	5 metric tons of TNT
5	179 metric tons of TNT
6	5,643 metric tons of TNT
7	179,100 metric tons of TNT
8	5,643,000 metric tons of TNT

Notable Earthquakes

Thought to be the worst earthquake in history occurred in 1960. The Chilean earthquake had a fault line of 621 miles (1,000 Km). The surface dipped 33 feet (10 meters). It was debated to have a 9.6 on the Richter scale. The deepest earthquake occurred on June 9. 1994 in the Amazon rainforest near northern Bolivia. That quake was measured at 8.9 on the Richter scale.

Year	Location	Richter Number
1811-1812	New Madrid earthquake, Missouri (series of quakes)	8.0-8.3
1899	Yakutat Bay earthquake, Alaska	8.3-8.6
1906	San Francisco earthquake, California	7.7 – 8.25
1960	Chilean earthquake, Chile	9.6

1964 Alaskan earthquake, Alaska 8.5

1971   San Fernando earthquake, Alaska 6.5

1976 Tangshan earthquake, China 8.2

1985 Mexico City earthquake, Mexico 8.1

1988 Armenian earthquake, Armenia 6.9

1989 Lorna Prieta earthquake, California 7.1

1990 Northwest Iran earthquake, Iran 7.7

1994 Bolivian earthquake, South America 8.3 (largest deep quake on record)

1994 Northridge earthquake, California 6.8

1995 Kobe earthquake, Japan 6.8

CHAPTER 5   GLACIERS

Glaciers have covered a major portion of the Earth several times during the Earth's long history. Scientists once believed that ice ages were the result of a shift of the Earth's orbit. Another belief was that volcanic activity once covered the Earth which blocked the Sun's energy. During certain times in Earth's past, global temperatures dropped between 7 and 21 degrees F. (4 and 12 C.) which caused 40 percent of the planet into a 100,000-year winter. These temperature drops plunged the Earth into as many as 20 ice ages in the past 2.5 million years. The most recent ice age retreated from North America, Northern Europe, and Northern Asia, about 11,000 years ago. The ice covered 16 million square miles (42 million square kilometers). Today, ice cover about one-tenth of all land and polar-regions. Ice covers 6 million square miles (15 million square kilometers). Today, ice covers 2 million square miles (16 million square kilometers) of Antarctica. Greenland and the northern polar areas maintain about 2 million square miles (6 million square kilometers) of ice. Seven to 8 million cubic miles of the Earth's water is in the form of ice.

Glaciers can be found today can still be found in several locations in the form of mountain glaciers and ice sheets. At one time the ice in Antarctica covered half of Eurasia in the east and from the North Pole to Kansas in the west. The ice sheet was nearly one mile thick and took thousands of years to melt. The retreating glaciers carved out the surface of the Earth creating deep cuts. Glaciers weighing trillions of tons dug out the area of the Great Lakes. The weight has depressed Antarctica a half mile.

Glaciers form where the average snowfall exceeds the melting and evaporation. Where the snow accumulates more than 100 feet, it compacts tightly in the bottom. It then melts slightly and refreezes. It takes a grainy form and then turns into solid ice. As more snow and ice is accumulated, it begins to spread outward. A mountain glacier will only appear in a single mountain valley. They also scrape up a large number of rocks and soil. As the ice melt, the rocks and soil are deposited. Large deposits of glacial sediment are called till. The wind-blown deposits are called loess. The till deposits create mounds and piles called moraines and string like, low-profiles valley ridges are called eskers.

Glacier Erosion – As glaciers move over the land, they create much erosion. Rocks embedded in the ice, together with the pressure of the ice mass, gouge out bedrock. When a glacier freezes to the bedrock, the moves on, it gouges chunks of rock.

Terrains of low relief – The land is sculptured by the ice sheets which had covered the land. Shallow basins for ponds are formed as a result.

Mountains within an ice sheet – The bedrock is mostly overridden and smoothed when the ice sheet was the thickest. Theses glaciers enlarge the valleys.

Mountains outside of the ice sheets – The land is sculptured by valley glaciers. They contain sharpened peaks, enlarged valleys, and basins.

Glacier path

Cirques – These are hollow at the valley heads where valley glaciers were originated. Erosion and weathering form a large amphitheater shape. The well-developed cirques have semicircular cliff forms with a basin at the base. The basin floor rises toward the entrance on the down glacier side. When the glacier melts, a pond may be the final result.

Artetes - are narrow, jagged ridges that form between cirques. They form as the headwall cliffs erode.

Cols – are low saddles between cirques that are formed by the intersection of curving cirques walls. They are often found at mountain passes.

Horns – are peaks made by the retreating of a cirque headwall from several sides.

Fiords – are glacial troughs with floors that are below sea level.

Truncated Spurs – are snubbed-off ends of ridges along sides of a valley.

U-Shaped Troughs – are valleys shaped by moving glaciers into troughs which assist the flow.

Rock basins – are hollows made by glacial erosion in the weaker rock.

Hanging Valleys – are tributary valleys of low gradient that quickly terminate into a steep wall of a large trough. Most often they are part of a waterfall.

Stairs – are found along the trough floor near the upper end of the main trough. They are at junctions with tributary troughs or sites of former icefalls.

Rock knobs – are resistant bedrock formations.

Rock drumlins – are similar to rock knobs, but have both front and rear smooth.

Striae – are grooves or scratches made by rock fragments in the ice. Some grooves could be deeper than one foot deep.

Crescentic fractures – are curved breaks in a nested series, made by embedded boulders.

Crescentic gouges – are depressions made by the removal of chips by vibrating boulders.

Glacial waste – Much of the rock waste is caused by glacial erosion. Most of the rock waste is deposited by the movement of the glacier and melting ice. The waste was deposited into the valleys and lowlands. Many rivers and valley were carved out by glacial erosion while some areas become filled with the rock waste. Glacial rock waste is called drift. The drift is sorted and stratified when it is deposited by water. The drift is unsorted and unstratified if it is deposited by ice activity. Unsorted drift is called till.

Terminal moraines – is till that has accumulated and pushed up when the glacier front is primarily stationary. Drift is accumulated in the crevasses near the front of the glacier.

Lateral moraines – are long, low ridges of wasted rock carried along the glacier's sides.

Eskers – are winding ridges of stratified sediments deposited by streams that ran with a glacier.

Kames – are irregular, dome-shaped, or rounded hillocks of stratified drift that was deposited by melting ice running off of a glacier.

Kettles - are depressions in drift caused by the melting of buried ice blacks.

Drumlins – is till molded by the ice sheet to form streamlined hills that are elongated toward the direction of the ice movement. They can be up to 500 feet high and several miles long.

Till plains – are deposits that completely bury pre-glacial landscape.

Erratics – are rock fragments that are transported by a glacier to a distant place and deposited as an isolated boulder.

Valley trains – are deposits of sorted drift extending down a valley from a moraine in a glacial trough.

Outwash plains – is sorted drift that is spread broadly as low alluvial fans bordering a moraine of a continental ice cap.

CHAPTER 6   STREAMS THAT SHAPE THE LANDSCAPE

Flowing water does more to shape the landform than any other geological activity. Streams have created valleys and have drained entire continents. Streams have designed floodplains and deltas and from large rivers to small streams. Water evaporation from land, rivers, lakes, oceans, is carried by the wind and falls back to Earth in the form of precipitation. That precipitation can be in the form of rain, snow, sleet, and fog. This precipitation falls to Earth and becomes runoff. It travels to streams, lakes, rivers, or to the sea. Some water settles into the ground and becomes ground water. It travels through the bedrock.

The groundwater filters down to where the rock and solid is saturated. The surface of this area is called the water table. In water bedrock areas, the water table tends to be irregular. Streams are usually fed by groundwater in the form of springs. A basin with a floor level below the water table will have a lake which will be fed by springs. The excess water is leaves by underground passages or through the lowest point in the local area. This is called the outlet.

Where the water supply is steady, permanent surface streams are available. They are usually from melting snow on a mountaintop or at a ground depression that is acting as a basin for the runoff. Occasionally it occurs from a spring on a slope.

Sedimentary Rock

Sediment are small pieces of rock carried by rivers and streams. They are broken rock fragments or weathered pieces of rock. They are deposited by water, glaciers, or wind. Geologists can find water, oil, or natural gas by examining where the rock grains came from and how they became to form layers at their current location. Sedimentary geologists use this information and the fact that sedimentary rock has spaces in which oil, water, and natural gas can become trapped in the rock.

The water current provides the energy to carry the sediment. When the water current begins to slow down, the heavier sediment is deposited to the bottom. The lighter sediment is carried until the force of the water current can no longer carry the particle. It's the force of the water current that determines where the grain is deposited. The shape of the streams morphology, or its path, affects the stream energy. If the stream has too many twists and turns, and too many obstacles to flow around, the energy dissipates and the larger particles are deposited to the bottom before the smaller particles.

Stream gradient is the calculation to determine how fast the stream is traveling. The speed of the water flow is determined by the stream's energy. Water rushing from the higher elevations is slowed down by dams, traps, or local rock formations. The less friction from these impediments would speed up the water flow. A waterfall would increase the energy and permit the grains of rock to flow faster and farther. The stream gradient is calculated by measuring the elevation of the stream to the physical distance of the stream. For instance, if the stream drops 10 feet for every 100 feet that it flows, the calculation would be a ratio of 1:10 or 1 to 10.

Notice the sediment and the water's edge. Is it a fast current?

Activity 1   How Do Sediments Settle From Suspension?

Note: Before you embark on an experiment or activity, be sure to wear eye protection, gloves, and proper clothing. Where required, obtain the necessary permission.

Strategy:

Step 1 – Using a piece of plywood or cardboard box, create a mountain slope. Use a yardstick or other item to reinforce the platform. Using a large plastic bag, cover the cardboard. Draw a mountain slope traveling from side to side and down the display. Using insulation foam, spray the edges of the 'stream' which should make a large 'S' shape.

Step 2 – After the foam dries, elevate one end of the mountain slope with a brick or other stable items. Allow the other end to rest on a flat surface.

Step 3 – Mix 1 tablespoon of sand, 1 tablespoon of dirt, and 2 tablespoons of gravel in 3 cups of water. Mix all of the ingredients until they are all suspended in the container of water.

Step 4 – Pour your mixture at the furrow located at the high end of your display. Write down your observations.

Step 5 – Raise the high end of your display by adding a second brick or another sturdy item. Repeat steps 3 and 4. Again, write down your observation.

Step 6 – Raise they upper end of your display four to eight times higher by adding more bricks or other sturdy support. Repeat steps 3 and 4. Write down your observation.

Observations:

1.   When you had stirred the dirt, sand, and gravel in the water, did the ingredients become suspended? After you

stopped stirring, what settled first? In a streambed, which would settle first?

2.   When you poured the ingredients down the mountain slope, which settled first?

3.   Where on the mountainside did the sand tend to settle, on the point bar or the cut bank? Where did the dirt settle?

4.   As the mountain grew higher when you raised the mountain display, what happened to the amount of sediment that settled on the mountainside?

Conclusions:

When searching for something in a river or stream, it is beneficial to know where to look. Heavier, larger items settle first. The sediment of all sizes eventually settles in a water current when the water flow slows down going around a point bar. Only the heavier sediment will settle in a cut bank area. These principles apply to any flowing body of water from streams to rivers.

Stream Channel – The strength of a streams energy is determined by the shape of the stream channel. Using a topographical map or aerial photograph, a geologist can determine the range of the stream gradient. The beds of streams, or the stream bottom, are channels. Channel banks may attach to valley walls or be separated from them by a floodplain. Channels contain sediments from mass wasting, weathering, and stream erosion. This waste is called load. Alluvium is when it is exposed by a shifting of the channel. Sediment is mainly silt and sand and some fragments from mass-wasted from valley walls. The mass wasting could be large boulders to sand, mud, and cobbles. The courser sediments known as bedload, move by jumping, rolling, bouncing, and scraping. This abrasion deepens the channel.

Valley development - Valleys are cut by streams eroding their channels. As sediments are pushed and dragged, erosion occurs by the impact and the suction of the loose material. As long as the erosion process dominates, the stream's channel is cut lower and lower. When that occurs, the valley deepens. Mass wasting on the valley walls and side-cutting, causes the valley to widen. The valley may be extended in the upstream direction by 'headward erosion'. This is the gullying by springs and slumping at the head.

Straight streams are observed when the direction of the stream flow is straight, without bends. Due to the surface landform and the flow from the higher elevation to a base, the force of the stream is likely to be rapid. The stream flows so rapidly that is erodes the channel regardless of the rock structure and surface material. These rapid streams are identified as having steep sides. The direction of the stream will show the direction of the higher elevation to the lower elevation as the stream flows downhill.

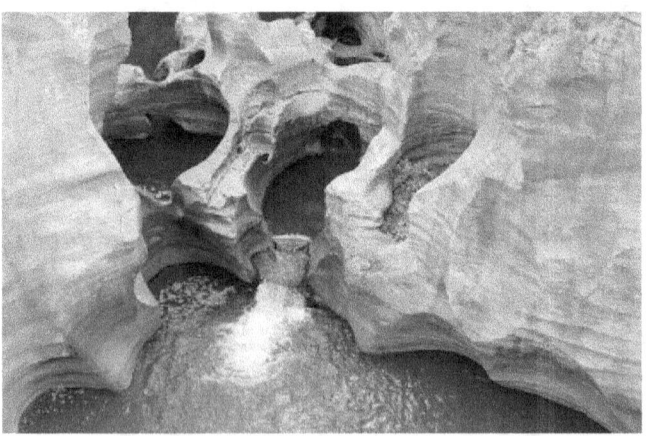

Pot holes

A young valley will have the rush of a fast moving body of water. The stream cuts down into the surface of the land making a deep ravine or gorge. This activity forms a deep ravine with waterfalls, rapids, plunge pools and potholes. When the stream matures, its' vitality is reduced and the valley bottom has lowered. Down-cutting has slowed or ceased. Waterfalls and side cutting have been eliminated. As the valley widens, streams swing from side to side. Sediment is then deposited to form a floodplain. These changes, along with other events such as climate change, valley drowning, and uplift widen the valley.

Meandering streams snake their way downhill. They meander from side to side as they flow toward the landform of least resistance. They are found in wide, flat areas. When the distance between the upper elevation and the lower elevation of a long distance, the energy slows down and the stream flows slowly. Meandering streams enforce less friction on the bank and channel bottom. These type of streams are often found near larger bodies of water such as lakes, oceans, or large rivers. Light sediment eventually fills the basin which slows the flow even more as the stream becomes shallow.

Dendritic streams resemble veins in a leaf as the stream spreads out into different directions. This type of stream contains both straight and meandering segments. These types of streams occur in hilly or mountainous areas. Water from one side of a ravine will meet up with water running from another side of the ravine as all of the segments of the stream flows to the lower elevation.

Trellis streams are unique in their form. This type of stream pattern is influenced entirely by the underlying rock formation. In areas where rocks have been thrust faulted, folded, or form parallel ridges, the streams run parallel with the base. These streams flow toward the gap that allows it to flow downhill to the next level. Trellis streams have high energy. Soft rocks erode easily and will erode between the harder layers of rock thus forming ridges.

Water gaps are valley segments that cross the parallel plain of ridges. A stream will cut through segments of eroded folds. When the stream cuts deep into the fold, it is called superposed. Other streams will flow parallel to the ridge and will eventually meet up with a straight stream that has found a gap in the ridge, or has created the cut on the ridge to flow downhill. The flow may have found a location where the ridge sags and the flow downhill begins at that point. Eventually, erosion of the gap will deepen the gap and allow the flow to move rapidly.

Wind gaps were previously water gaps that were converted by stream capture. This occurs when one stream by head-ward erosion intersects another stream of lower gradient and diverts the water into its own channel. The captured stream's channel was a water gap is now a wind gap. The valleys on either side of the ridge at opposite ends of the gap are cut deeper below the floor of the gap. The wind gap may appear as a shallow notch in the ridge.

Lakes – A lake is a wide part of a stream. A stream fills the lake at one end and the overflow spreads over the low area. The stream continues to flow downhill. Lakes are still water for the most part because the flow is gradual. Because the flow is slow, sediment falls to the bottom of the lake and the lightest sediment remains suspended. Algae and other lake life forms and clouds the lake. Lakes are usually short-lived because they fill with sediment and will empty as their outlet streams cut down. Wave-cut cliffs and terraces may form along large lakes as windblown waves cut through the soft rock. The Great Lakes show features similar to that of an ocean shoreline.

Activity 2   Reading Stream Gradient on Topographic Maps

Know how to recognize the four types of stream formations: straight, dendritic, meandering, and trellis. Demonstrate your ability to read a topographical map by calculating the stream gradient.

Strategy:

Step 1 – Using a contour map, identify the types of streams found on the map.

Step 2 – Find the starting and ending point on the map where the stream crosses a contour line. Measure the distance on the map between the points with a ruler and convert it to feet or miles using the scale on the map.

Step 3 – Read the elevation of the stream from the beginning point to the ending point from the previous step. The interval in the contour line is the elevation. Use this information to find the elevation of the contour lines that cross the stream.

Step 4 – Calculate the stream gradient. That is the ratio between the elevation and the distance. Stream gradients are expressed in ratio form. Reduce the ratio to the lowest common denominator.

Step 5 – Repeat steps 1 through 4 using other topographical maps.

Observations:

1. Which type of stream flows the fastest and why?

2. Which type of stream flows the slowest and why?

3. Which stream would carry the largest grains of sediment and why?

4. What type of stream gradient would a waterfall have? Do waterfalls flow fast or slowly?

Conclusions:

Straight streams have the most energy and highest gradient and carry the most sediment. Meandering streams have the least energy and they carry the least sediment and smallest sediment.

Stream Landforms

Assume that you were a prospector in the California gold rush in the 1800's. You would have carefully observed the stream landforms while you pan for gold.

Cut Bank – The S-shaped curves of a meandering stream you will observe that the force of water pushes on the outside of the bend. The outside of the bend is continuously eroded by the rushing water. The outside of the bend is called the 'cut bank'. If you ever notice in a car race, the driver on the outside of the bend has to drive faster to keep up with the inside lane drivers. The outside of a meandering stream has to travel faster than the water on the inside bend. Because the outside bank water has to travel faster, it picks up sediment and erodes the cut bank. No sediment grains will settle in the cut bank.

Fill Bank and Point Bar – The water on the inside of the bend slows down and has low stream energy. The larger grains of sediment are dropped in this area. The inside of a river bank bend is called a 'fill bank' or 'point bank'.

Delta – Where a stream flows into an ocean or lake, sediment is deposited and a delta is formed. Delta's can have various patterns, but they all have a point where the delta ends and the slope drops rapidly into deep water. Stream channels occur within a delta. Where the energy of the stream drops to zero, grains of sediment are deposited.

Medial Channel Bar – Sometimes called a 'Sandbar' a medial channel bar is an elongated mound of sediment in the middle of a channel or waterway. A delta may have a number of sandbars.

Activity 3   Using a Magnifying Glass, Find Sediment

Using a magnifying glass, observe stream sediment. Some sediment is so small that it doesn't settle to the bottom right away. Take a look at the suspended sediment.

Strategy:

Step 1 – With the proper permissions, and a friend or two, visit a nearby stream. Using a clear plastic cup, scoop up water in the stream in an area where the stream energy is slow. With a magnifying glass, examine the cup of water carefully for suspended sediment. Write down your observations.

Step 2 – Find a fill bank location at another point along the stream. Scoop up a second sample of stream water. With your magnifying glass, observe the sediment and write down your observation.

Step 3 – If you have access to a microscope, place a drop of your stream water on a slide. Carefully observe your sample. Write or draw your observation.

Observation:

1.   Did you see more sediment with your magnifying glass than without the magnifying glass?

2.   Was there more sediment in the first or second sample?

3.   What did you observe through the microscope?

4.   Do you think that sediment impacts the stream bottom? Why?

Conclusions:

Much of the sediment suspended in stream water is too small to see without magnification. Energy from flowing water can move everything from boulders to microscopic sediment. Our planet's landscape is constantly changing due to flowing water.

Activity 4   Water Direction

If you ever walked across a dried river bed, you will find clues to indicate which direction the river flowed.

Strategy:

Step 1 – With permissions and with a friend or two visit a nearby stream. Observe the direction of water flow. Drop a leaf or stick that will float to confirm the direction of water flow. If the stream is dried up, look for clues that indicate which direction that the stream water was flowing.

Step 2 – Search for a second stream that is feeding into your current stream.

Step 3 – Search for an obstruction in the stream. It could be a rock or tree. The upstream side may have sediment buildup while the downstream side may be hollowed out.

Step 4 – Search for debris such as leaves and twigs wrapped around rocks and trees along the bank.

Step 5 – Search of grass, reeds, or litter bending downstream.

Step 6 – Record all of your findings in notes and drawings on paper and/or take snapshots of the findings. Share your findings.

Observations:

1   Were you able to see the sediment carried by the stream current?

2   Did the second stream and first stream form a 'V' where they met? Did the 'V' point upstream or downstream?

3   What type of obstruction did you find in the stream? What had accumulated on the upstream side of the obstruction? Was the obstruction hollowed out on the downstream side?

4   Did you find leaves and twigs wrapped around rocks and trees? Did the height of the debris indicate the height of the stream once overflowed its banks?

Conclusions:

Clues of the flow of water can be found even if the stream is dried up. When to streams meet, they form a 'V' pointing downstream. Obstructions such as rocks and trees collect debris such as leaves and twigs. All streams leave water flow direction clues.

Features of Valleys

Potholes – are smooth, circular, and deep in the bedrock. They are created by eddies and abrasive sediment. They take decades to form and are about a yard deep.

Waterfalls – They occur at young valleys. A cliff with a rock mass that is stronger than the rock mass below it. This occurs during glaciation or from faulting, a lava flow, landslide, or valley-blocking. A series of falls is called a cascade. Waterfalls eventually subside due to an undermining base, abrasion, or a change at the top of the falls.

Rapids – are areas of rushing water sometimes referred to as 'white water' which forms around rocks protruding from the stream bed. The rocks may be weathered valley walls or edges of strata. They are often found near volcanic areas or faulting. When the valley bottom begins to become graded by erosion, the rapids begin to disappear.

Plunge pools – are bodies of water at the base of waterfalls in basins caused by erosion from the falls. The sediment at the base of the falls causes abrasion from the constant churning of the sediment.

V-Shaped valley – is formed where there is mass wasting on walls in relation to stream erosion. The loose mass wasting forms an angle about 35 degrees or less.

Undercut slope – is a steep cliff or slope outside of a river bend. Overhangs form as a result of side cutting by the streams as it rounds a curve. Eventually, the overhangs fall forming a steep wall.

Slip-off slope – is a gradual slope on the inside of a river bend, opposite of an undercut slope. Side cutting is minimal.

Vertical walls – is found at valley walls where little mass wasting or side cutting occurs. In areas where the channel is straight, a stream will make a slot-like cut. At bends in the rock where weathering and mas wasting does not occur and does not keep pace with side cutting, then overhangs may develop which would form niches or alcoves.

Walls of unequal slope – are walls that are affected by stream erosion, mass wasting, and weathering which is different on opposite sides of a valley.

Ribs – are edges of horizontal strata projecting from valley walls. This is common among areas of sedimentary rock.

Benches – are rock platforms that extend out from the base of the scarp or cliff. They are found on a valley wall toward a stream channel that has been cut to a lower level. Typically, a bench is found on a single strong layer overlaid by a weaker layer.

Natural bridges – are rock arches. Often the arch is made by a stream cutting through a divide. Others are made by a stream-made tunnel. Some are parts of a cave rook that had collapsed.

Towers and pinnacles – are a result of predominantly vertical jointing. The joints define the form while weathering and erosion isolate them from each other and from the valley wall. Some are hardened curves of volcanic rock.

Buttresses – are rectangular rock masses protruding out from valley walls. They may resemble buttresses from a Gothic church. It is the sedimentary rock with wide rectangular jointing.

Deposit Features

A stream has the ability to carry solid particles (sediment) depending on its velocity, volume, density, amount of matter, and the type of particles. If the sediment is too heavy for the stream to carry the particles further, it is then deposited. The finer material remains suspended in the water until there is a loss of turbulence. The deposit of sediment occurs anywhere along the stream. It can range from river stone to fine dust particles. The sediment becomes alluvium when it becomes isolated by a shift of the channel.

Alluvial cones and fans – are masses of sediment deposited at the foot of a mountain ravine where a stream quickly reaches the valley floor. Depending on the change of the gradient, the sediment is sorted from fine to coarse material. Alluvial cones are found at the bottom of the steep slopes. Alluvial fans are found at the bottom of gentle slopes and are larger than cones.

Alluvial terraces – are level surfaces on deposits extending out from a valley wall. A terrace indicates a former level of a stream deposit. A change in the conditions of a flow caused the stream to regrade itself where it smoothes the bed at the lower level. This act leaves the former deposit level above the stream. This is common in a valley where the streams were formerly swollen with rock waste and melting glaciers.

Braiding – is the division of a stream into channels that are separated by elongated islands of sediment. When a stream slows, it deposits is sediment and widens. This causes the channel to widen, and the channel is then clogged with sediment.

Floodplains – are horizontal surfaces on fine-grained deposits left by floods. The flood waters overflow from a curved channel that is no longer able to cut down into a bank. The floodplain starts with the deposition of sediment as point bars at the base of slip-off slopes while the channel is shifting laterally around the bends.

Meanders – are smooth, loop-shaped curves developed by continued cutting and filling in the channel. Meanders migrate downstream while spurs are eliminated.

Cutoffs – are short channels across necks of meander curves. When a curve forms a near circle, the neck tends to narrow. The flood will cross the neck and narrow. This forms a shorter channel.

Oxbow lakes – are formed in a cut-off meander loop. These lakes are fed by seepage into the lake.

Meander scars – are parts of meander streams that are filled with sediment. When scars are in a series, the evidence of how a stream repeats a cycle is exposed. The scar shows how the stream was cut off and how loops were filled while making new channels downstream.

Natural Levees – are curved ridges flanking meander streams. They are usually 10-12 feet above the floodplain. They form when flood waters overflow the channel. Sediment is then deposited parallel to the channel. The stream builds up its bed between the levees. When the stream level rises above the levee, the levee breaks forging water with sufficient energy.

Deltas – are accumulations of sediment that is deposited where streams flow into lakes or oceans. Deltas in estuaries or lakes are long and narrow. During a flood, a delta receives more sediment. The sediment is rich in organic material.

Drainage systems – are combined groups of valleys where running water on land moves toward lower levels. When a new land surface forms valleys will begin developing on the slopes. Through erosion, they link with other valleys forming a drainage system. The drainage system patterns depend on the surface rock structures. This feature can be traced on contour maps.

Topographical Maps

Topographical maps are used by geologists to determine the three dimensional Earth from a two-dimensional flat piece of paper. Using contour lines, a geologist can determine land surface. The elevation is the vertical height above sea level. Areas of the same elevation are displayed by connected contour lines. Where the lines are close together, that location is steep. Where the lines are farther apart, those areas are more level indicating gradual slopes.

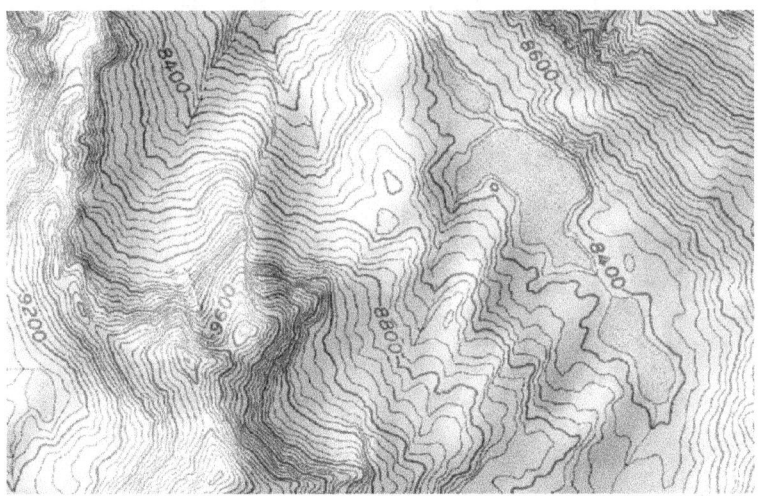

CHAPTER 7   ENERGY

The world economies depend on energy for manufacturing, business, and the transportation of goods and services. One-half of a trillion dollars is budgeted for energy in the United States alone. Almost all countries on Earth depend on energy for commerce.

Natural Resources

Before natural resources can be used, they must be found, extracted, and processed. Throughout history, natural resources have been used for the daily uses of civilization. Rock is crushed and used in building material. Ore is mined and processed for metals. Oil and natural gas are used for plastics, medicine, and transportation. Coal is used for heat and electricity. Jewelry comes from gemstones. Everything man uses comes from the Earth.

Finding natural resources can be obtained from a knowledge of the Earth's surface to exploring deep into the Earth's crust. Obtaining the natural resources can be a simple process of digging sand from a river bed. Obtaining resources can be difficult such as drilling for oil in the oceans or blasting tunnels in rock to obtain ore. Another feature to consider is the transportation of the natural resource. It can be as simple as loading a truck or train to a more difficult project such as laying a pipeline many miles in length. In each situation, the cost must be considered. It is up to the geologist to find the natural resources needed by the people of Earth.

Electricity

Around the world, electricity is used to provide heat and power machinery, appliances, and electronics. Electricity is clean in that no waste byproduct is left behind. In the United States, only 21 percent comes from nuclear energy, while 51 percent of electrical generation comes from the burning of coal. Natural gas provides 17 percent while hydroelectric facilities only provide 6 percent.

Anthracite Coal

Fuel oil provides 3 percent and all other electric generation provides 2 percent. Although solar and wind power generation is new, we still depend mostly on coal to generate our electric needs.

Lignite Coal

Coal – Between shale and sandstone are layers of sedimentary rock containing coal. Coal is formed when plants form in swamps and bogs and are buried under sediment. Oxygen is necessary for decay which doesn't occur because the material is sealed. The decay of coal depends on pressure, heat, and length of time that the coal was buried. The first stage of coal formation is called 'peat'. When peat is dried, it can be burned and used as a source of heat and fuel.

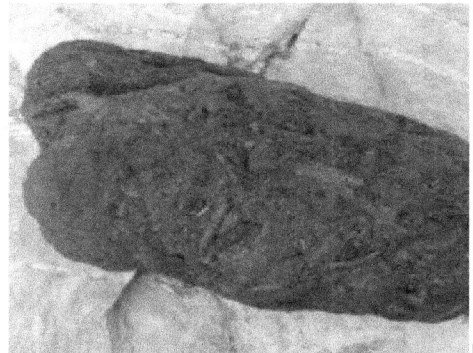
Dried Peat

As time moves on and with a deeper burial, water is extracted from peat. This second stage of coal development is called 'lignite' or brown coal. As time, pressure, and decay increases, the third level of coal formation occurs. The level of carbon is increased and lignite becomes bituminous coal which is the most widely used form of coal. When bituminous coal is folded in its formation, the highest grade of coal is formed called 'anthracite'.

Peacock coal

Oil and Gas drilling in Alaska

Oil and Natural Gas – All living things contain carbon and hydrogen atoms. Oil and Natural Gas contain the same carbon and hydrogen atoms. Scientists believe that microscopic organisms and plants were preserved on the muddy ocean bottoms. They were rapidly covered with mud so that the oxygen could not decay the plants and animals. The material was covered with sand and clay. This sediment formed rock filed with pores containing the clay, water, and organic material. While they were buried, pressure, bacteria, and heat caused chemical changes in the material. The carbon-based molecules became oil and natural gas. The oil and natural gas are lighter than water so it will tend to rise upward. The oil and natural gas rise until it reaches a trap. This would be a layer of nonporous rock. Occasionally, the trap will cover a large area and the accumulated oil and natural gas forms an oil and gas reservoir. Rock layers bend upward in this process. Geologist search these areas called 'anticlines' for evidence of porous rock sealed by impermeable rock. Core samples are then performed and examined for evidence of oil and natural gas.

Shale

Shale oil – is a rock that had formed by plant and animal matter with ocean bottom mud. This matter compacted with mud is called 'kerogen'. All oil, gas, coal, and kerogen are hydrocarbons which are composed of carbon and hydrogen atoms. Oil shale is mined as any other rock. When it is heated, it yields the kerogen. The cost of mining and processing shale oil is very high.

Nuclear energy – Uranium is used for heat in order to boil water which turns the turbines of a generator. This is how electricity in a nuclear power plant uses uranium. Uranium is a heavy metal found in the mineral, uranium. This can also be found in carnotite and autinite. These minerals are radioactive.

Hydropower – Dams produce free renewable energy that is free of pollution. The turbine wheels turn with the weight and pressure of the water flow. The wheel is attached to a generator. The dam also provides water for communities and agriculture. The risk of floods and droughts are reduced. However, the dam prevents fertile sediment from flowing downstream. Another problem with dams is the evaporation issue which could alter the local climate.

Wind – Large windmills are turned by the wind in order to generate electricity. Many countries now use windmill farms to assist with their electric power needs. In a previous time, windmills were used to pump water to livestock and irrigate farmland. The downfall of a windmill farm is that it requires a large area of land and perpetual wind.

Tidal Power – In a few areas where the tide rises and falls, tidal power is used. As the tide drops, the water in a secured area is slowly released to turn turbines that will generate electricity. The downside is that tidal power requires a large coastal area and the shoreline ecosystem may be damaged.

Solar energy – Today, solar energy is used to heat water for homes and businesses. The scientist is improving the efficiency of the solar panels. NASA uses solar panels for its spacecraft. The downside is that it requires a large area and that area must have abundant sunny days.

Geothermal energy – This type of energy can be obtained in areas where magma is near the surface. Pipes are drilled into the surface and place near the hot magma. This heats water which is converted into steam. The steam turns the turbines and electrical energy is produced.

Geologist analyze mud core samples

How are oil and gas explored? Geologists use high-tech tools to search for oil and gas. They produce a map of what to expect to find under the surface. This primary map is a mere expectation.

Observing rock layers – Water deposits sediment which accumulates in basins. Geologists study and record the basins stratigraphy. These are the layers of rock beneath the surface which is called the 'basement'. Geologists know that if certain layers of rock hold a reservoir of oil in one part of the basin, they can assume that oil could be found in other parts of the basin. If a certain rock can be found 6,000 feet under the surface of the basin, then it can also be found 6,000 feet under a different part of the basin. Locating the rock layer formation in a different part of the basin is called the 'stratigraphic correlation'. Geologist use tools such as electrical well-logs, core samples, reflection seismic instruments, and cuttings. Reflection seismic is a process of sending acoustic waves (sound waves) deep into the Earth using a wave energy source. The mechanism will either use a hammer type device, but most often uses a small explosive charge. Since sound travels faster in solid objects than in the air, the reflected waves are analyzed to form an image of what is below the surface. The instrument can also determine the type of rock layers, such as limestone or sandstone that the acoustic waves had passed through. Computers measure the time it takes for the reflected wave to travel round trip. The graphic displays can identify where oil and natural gas is located.

Electric well-logs – are electrical instruments placed in a well after it has been drilled. The object of this instrument is to create a record of the stratigraphy and rock properties it encounters in the well. Instead of a sound wave, an electric current is passed through the rock formation. This long cylinder instrument is called a 'sonde'. This tool is connected by wire to a computer on the surface. The sonde measures the electric current passes through the rock formation. Comparing the electrical conductivity to rock properties in other parts of the basin, geologists predict the rock formations in other parts of the basin. Engineers then plan how to drill for oil and gas.

Core samples – Usually, well logs and seismic readings render enough information to begin drilling a new well. On some occasions, engineers and geologist test a core sample. A core sample looks like a cylinder shaped rock. Geologists can determine the rock's porosity, which is the space between the solid grains. The core sample is obtained with a tool that has a hollow drill bit that is a cylinder in shape. The drill cuts a cylinder of rock which is brought to the surface for examination. The ground rock material is called 'cuttings'. Although the core is 1 to 2 ½ inches in diameter, the sample can be hundreds or thousands of feet long.

EXPERIMENT: Make a Core Sample.

- Cut the top off of a clear two-liter plastic bottle

- Mix equal parts of dry plaster and sand.

- Fill the bottom 3 inches of this mix.

- Mix equal parts of plaster and dirt.

- Make a second layer of dirt mix.

- Place leaves, shells, and twigs between each layer.

- Make a third layer of plaster, gravel, and rocks. Fill the bottle.

- Fill the bottle with water. The next day, remove the core sample.

Creating an Exploration Map

Geologists first use a seismic survey to get a picture of the subsurface structures. The structures can include tops of formations or other underground features, such as caves. With the information accumulated, geologist and geophysicist analyze the basin stratigraphy on a seismic print. Using other information from previous explorations, geologists will use an electric well log to plot the depth of oil or natural gas. A subsurface structure map using the depth-to-formation numbers from the well-log plots are used to determine the bottom formation and thickness. The subsurface isopach map will show the expected thickness of the target reservoir.

Activity 5: Drawing a Subsurface Structure Map

If you've ever drawn an image by connecting the dots on a puzzle, you will be able to draw a subsurface structure map. However, this image maps the buried top of a target rock formation.

The elevation used on most subsurface structure maps are shown in negative numbers. These numbers indicated the values below sea level. Sea level is a 'datum' which is the standard reference position in all wells. The elevation point at sea level is zero.

If you have access to well data or electric well-logs, use that information to create your map. Otherwise, use the data in the chart below to build your map.

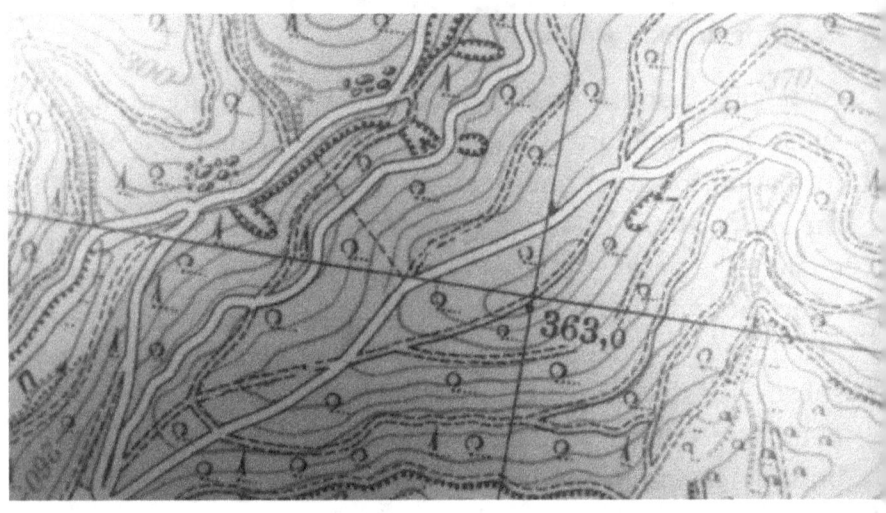

Topographical Map

Well Number   Depth to Structure

	(Subsea level) Feet	+6,200 Feet
1	-6,790	580
2	-6,810	610
3	-6,840	640
4	-6,435	235
5	-6,241	41
6	-6,438	238
7	-6,294	94
8	-6,489	289
9	-6,750	550
10	-6,555	355
11	-6,750	550
12	-6,780	580
13	-6,805	605

14	-6,350	150
15	-6,395	195
16	-6,428	228
17	-6,463	263
18	-6,658	458
19	-6,679	479
20	-6,742	542

Strategy:

Step 1 – The numbers are negative in the depth-to-structure because they are below sea level. To simplify the map, add 6,200 feet to each depth in the blank column.

Step 2 – Draw, trace, or copy the well location map and label each well with the depth below 6,200 feet that you have calculated in step 1.

Step 3 – Notice the elevations. Decide whether to draw the contour lines on 50-foot intervals. Find the lowest point.

Step 4 – Suggestion: Use a light pencil or dashes until after you have checked the points in step 7. Draw a line between the four lowest points for the -600-foot contour line.

Step 5 – Draw the contour line for -550 feet. Does this line intersect points?

Step 6 – Continue sketching contour lines between points until you reach the highest elevation. Note: Contour lines cannot cross one another.

Step 7 – Check the contour line that you have drawn. Re-check your lines again. Be sure, for instance, that an elevation of -289 doesn't fall between the -300 and the -350 contour lines.

Step 8 – Search for a structural high, the location of the highest relative elevation. It resembles a hilltop of a topographic map, but actually, it is the structurally highest point on the map. Oil and gas will migrate upward since they are lighter than water. The oil and gas will rise through the rock and collect in a structural high. Label your map where you would expect to find an oil reservoir to exist.

Observations:

1. Is the lowest point well 3 or 5? Why?

2. Are some contour lines spaced out more than others? Where is the steepest section of the formation?

Conclusions:

Three-dimensional surfaces can be clearly seen on a contour map. Drawing a contour map reveals hills and valleys that would be difficult to find by reading the well data. A subsurface map is an important geological tool.

Isopach map – Geologist make a structure map of the bottom of the formation. Then they calculate the thickness by subtracting the depth of the top from the depth of the bottom. The map showing i=formation thickness is called an isopach map and can assist the geologist or engineer to determine whether there is enough rock formation present to make a sufficient reservoir for oil or gas.

Offshore drilling – Oil companies drill on dry land and in deep water. Offshore drilling is much more expensive for the oil companies. Some offshore drilling rigs float and drill deep into the ocean floor. When an oil company drills from a single platform, they may establish nearby platforms which all drill from the same hole.

Activity 6   Extraction Tabletop Display

The geologist has located a large oil and gas reservoir. Now, what happens? For this activity, make a tabletop display and explain either oil and gas extraction or coal extraction. Share your display with a group of people or a badge counselor.

Strategy:

Step 1 – Choose oil and gas OR coal as your topic. Make a three-panel display board. Place one large label in the center that reads either "Oil and Gas" or "Coal." The three labels for the three panels could be labeled "Exploration," or "Extraction," and "Processing".

Step 2 – With the proper permissions and a librarian's help, search the public library databases for articles on oil and gas OR coal. Explore the internet using a search engine for terms like "oil exploration: OR "coal mining", or other similar search terms.

Step 3 – Decorate your display board with interesting facts you have found on extraction.

Step 4 – Give a five-minute presentation on your findings to a group or a badge counselor.

Observations:

1.   Are geologists more involved in exploring for oil, gas, and coal or processing it?

2.   What have you learned during this project that you didn't know before?

3.   Are there oil, gas, or oil reserves in your state? What states do have oil, gas, or coal reserves?

4. Does your home use natural gas? If yes, in what ways is gas used in your home?

Conclusions:

Now there is an understanding that when gas is pumped into a car, or lights are turned on in your home, many people, including geologists, have worked to find, extract, and process the oil, gas, and coal.

Activity 7 An Operating Drilling Rig Experience

What would it be like to work on an oil drilling rig as a geologist?

Strategy:

Step 1 – Locate a geologist that has worked on a drilling rig. You may be able to locate the drilling rig geologist using the address at the rear of the book, or searching organizations online. With the proper permissions, contact the geologist via email or phone. Explain your interest in geology. Some museums may have similar displays. Contact museums in your area.

Step 2 – Where feasible, arrange to visit the geologist at work. Ask to see core samples.

Observations:

1. What is drilling mud and what purpose does it serve? What are drill bits and drill pipe?

2. How many barrels of oil does the geologist think the reservoir holds? How many gallons of oil is that? (There are 42 gallons of oil in a barrel) How much of the oil is expected to be recoverable?

3. Ask the geologist what is the most satisfying part of working on a drilling rig. Find out about obtaining a geology degree. Ask the geologist why he selected this career path?

Conclusion:

Viewing life of a geologist on a drilling rig will give you a look at a geologist daily life regarding that experience.

Rocks are a mixture of one or more minerals. Rocks, such as marble contain only one mineral, calcite. A crystalline substance that is a naturally occurring on Earth is considered to be a mineral. The mineral must be solid and homogeneous which means that the mineral must be of the same material throughout. The mineral must have a distinct set of chemical and physical properties. A diamond is the hardest mineral known to man. It occurs naturally on our planet.

Minerals

The most common minerals on Earth are calcite, quartz, mica, feldspar, gypsum, and pyrite. Geologists know that the most common minerals are composed of the most common elements. The most abundant element on Earth is Oxygen and silicon. The less common minerals are usually metallic in nature and are silvery gray and cubic galena. In earlier times, mineralogists used comparative scales to identify minerals and their physical properties.

Elements found on the Earth's crust:

- Oxygen   46.6%
- Silicon   27.7%
- Aluminum   8.1%
- Iron   5.0%
- Calcium   3.6%
- Sodium   2.8%
- Potassium   2.6%
- Magnesium   2.1%
- All other   1.5%

The comparative scale singles out the unique qualities of the mineral. Color may be a distinctive trait of the mineral. Minerals may be identified by using two or three comparative scales on each specimen. There are eight physical properties to compare: hardness, specific gravity (mass), color, cleavage, fracture, luster, crystal form, and streak.

Hardness – The test for hardness is a 'scratch' test. Simply a scratch on mineral against another mineral. The harder mineral will scratch the softer mineral. Mineralogists have developed a numerical scale using the relative hardness of common minerals. It is called the Mohs' scale. When chalk is used on a blackboard, the chalk wears down, not the blackboard. Therefore, the blackboard is the harder of the two.

Mohs' Comparative Scale of Relative Mineral Hardness			
Mineral	Scale of Hardness	Common Use of Mineral	Common Minerals with Similar Hardness
Talc	1	Talcum Powder	
Gypsum	2	Plaster of Paris	Fingernail 2.5
Calcite	3	Cement	Copper penny 3.5 pre1973
Fluorite	4	Fluoride in toothpaste	
Apatite	5	Fertilizer	Steel nail 5
(Plate Glass)	5.5	Glass products	
Orthoclase	6	Artificial teeth	Knife blade 5.5
(Steel file)	6.5	A Tool used to sharpen	
Quartz	7	Quartz watch/Quartz sand in porcelain	Glass 6 to 7 / Streak plate 6.5 to 7 / Hardened steel file 7+
Topaz	8		
Corundum (ruby/sapphire)	9	Ruby or sapphire for jewelry or lasers	
Diamond	10	Jewelry or cutting tools	

Look at Quartz on the Mohs' scale. Quartz hardness is a 7. Quartz can scratch everything less than a 7. Topaz with a hardness scale of 8 can scratch Quartz. A diamond with a hardness scale of 10 can scratch anything.

Specific Gravity or mass – is the weight of a substance compared with the weight of an equal amount of water. The mass of lead will have a high specific gravity. Aluminum, a lighter material, will have lower mass or specific gravity. Geologist sort minerals by mass.

Color – The most common, and obvious feature of minerals, is color. However, the color is the least reliable as a mineral test. If a mineral contains impurities, the color could be different from other samples of the same mineral. Geologists use color as their first pass comparison before using other tests.

Fracture and Cleavage – If struck against a hard object, the mineral is said to 'fracture' if it breaks and exposes an uneven surface. Types of fractures include uneven, hackly, or conchoidal. If the mineral breaks along a smooth surface and exposes a regular form, it is said to be 'cleave'. Smoother surfaces are evidence that the minerals have been chemically bonded. Cleavage is an identifier for mineral identification. Fracture is an indicator that there is a lack of cleavage.

Luster – is the appearance of the mineral in normal light. If the mineral gives off a glassy appearance it is considered to have a glassy luster. If the mineral gives off a dull or dirty appearance, it is said to have an earthy luster. Other labels of the luster identifier are oily, metallic, silky, or waxy.

Crystal form – When a mineral cools slowly, it has the ability to grow into a larger crystal. Geologists can identify minerals by their characteristic crystal shape. For example, a quartz crystal may form in a volcanic steam vent. If the quartz is left without any constraint, the crystal will grow one molecule at a time until the crystal has enlarged. If the quartz crystal was permitted to cool rapidly, any crystal formation would be very small. Granite forming underground contains quartz but is constrained by the compression of other mass amounts of rock. The crystals are very small in granite.

Streak – Geologists use a piece of rough porcelain for a streak test. The streak test is the process of rubbing a mineral against the porcelain and observing the powdery residue that is left. The porcelain has a hardness of 7. The porcelain is called a 'streak plate'. Sometimes the residue left on the streak plate can be deceiving. For instance, the iron mineral called limonite is yellow but leaves a streak residue of brown which is common for any iron mineral.

Striations – are the narrow lines or bands that run across various mineral surfaces.

Rocks

A rock is a naturally occurring, consolidated solid substance composed of a mineral or a mixture of minerals that form a part of the Earth's crust. The Earth's crust has three classes of rock based on their origin: igneous, sedimentary, and metamorphic. Of the 2,000 varieties of minerals recognized in rock, only a few forms the bulk of all rocks. Each rock contains two or three types of minerals. Rock varieties vary from region to region. The kinds of minerals that form, and the ways in which the minerals are arranged, strongly influence the nature of the way landforms are shaped by erosion.

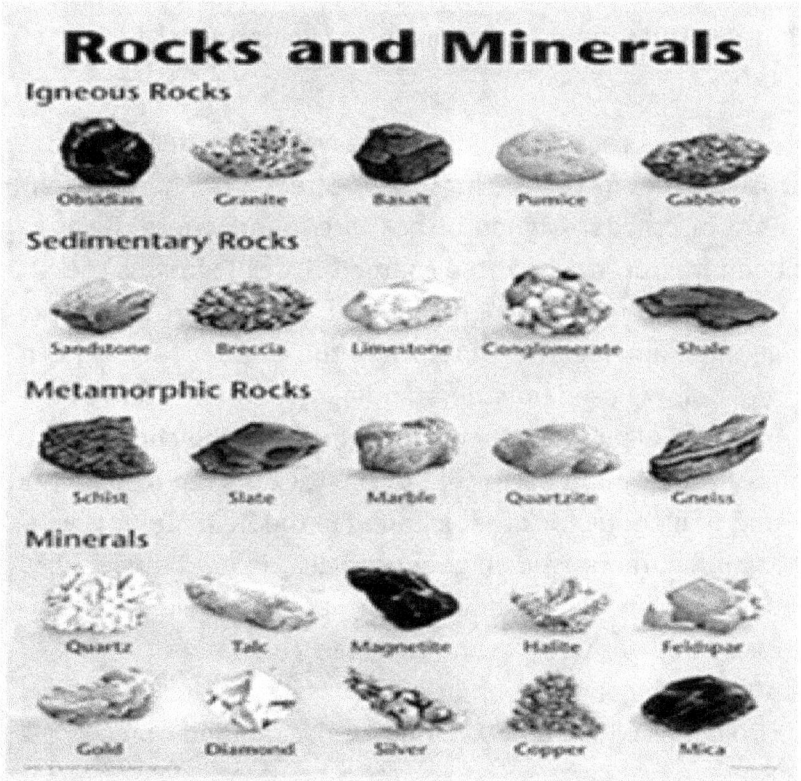

Igneous rocks – make up 95 percent of the Earth's crust. They form from a molten, or partially molten material called magma. When magma reaches the surface, it is called 'lava'. Igneous rocks are classified on the basis of texture: course-grained, fine-grained, or glassy and mineral composition. There are two basic types of igneous rocks: intrusive and extrusive. Igneous rocks form by cooling and solidification of magma 'in' the crust. This type of rock is cooled deeper inside the Earth and is called 'intrusive'. Intrusive igneous rocks are hard and heavy. Intrusive igneous rocks are coarse-grained and have mineral crystals or grains the size of a fingernail. Sometimes igneous rocks called 'porphyry' form. Porphyries show two stages of cooling with large crystals called phenocrysts embedded in a fine-grained groundmass. Grain size, or rock texture, provides information about the depth at which the minerals in the magma began to cool, forming the igneous rock. These deep rocks rise to the surface when tectonic forces uplift the area and the material covering the intrusive igneous rock has eroded.

Igneous rocks formed 'on' the crust are referred to as 'extrusive' igneous rocks. They may be heavy and hard to light, crumbly, or powdery. Igneous rocks are found where the crust has been fractured. The three characteristic of igneous rocks are texture, color, and mineral content. The texture shows the cooling time and location or depth of the igneous rock when it formed. Color is used in understanding the composition of the magma from which the igneous rock formed. The chemical composition or mineral content of the rock is used to classify the kind of igneous rock that is found in the field. Color is an important characteristic of igneous rocks. Lighter colored igneous rocks such as granite are formed from sialic magma. They are filled with quartz and feldspar. The darker colored igneous rocks such as green or black basalts are formed from mafic magmas which a filled with magnesium and iron.

Granite – is intrusive and has visible interlocking crystalline grains that are mostly feldspar and quartz. Granite often forms mountain cores.

Gabbro – is intrusive and has visible interlocking crystalline grains that are mostly plagioclase feldspar and ferromagnesion minerals. Hard as steel and dark in color. Found in humid areas and in geological dikes and sills.

Felsite – is extrusive and has microscopic interlocking crystalline grains that are mostly orthoclase feldspar, quartz, mica, and ferromagnesians. It is hard as steel and light in color. Occurs as lava flows.

Basalt – is extrusive and has microscopic interlocking crystalline grains. It is hard as steel and is dark green to blue-gray in color but weathers brownish. Occurs as lava flows and may fracture due to contraction and solidifying. It has small pores that are filled with other minerals.

Andesite – extrusive and has mixed microscopic and visible interlocking grains. It is as hard as steel. Andesite has medium shades of red, green, and gray.

Sedimentary rocks – make up most of the top layers of the Earth's crust. This is because it is where weathering, erosion, and deposition occurs. Sedimentary rocks cover about 75 percent of the world's land surface but are only 5 percent of the volume of the crust's outer 10 miles (16 kilometers). Sedimentary rocks are hardened layers of sediment that form when preexisting igneous, sedimentary, and metamorphic rocks that are physically, chemically, biochemically weathered or eroded. This sediment is redeposited by natural elements such as ice, wind, and water. Sedimentary rocks are classified as detrital. That is having particles of different size and shape that is naturally compressed and cemented together chemically. They are crystallized precipitates from sea water and form in layers.

Limestone – is mostly calcite or dolomite. The crystals are of organic shells, precipitates, and mechanical origin. Limestone is white to blue in color and has a hardness equal to copper. In humid climates, limestone could be found in sinks, caves, and valleys.

Shale – is primarily clay minerals and quartz that is colored by carbon, metallic oxides, and ferromagnesians. Shale is fine-grained, thin-bedded, and flaky and is almost any color especially red, brown, or black. Shale is soft unless it is infused with silica. Shale can be found in valleys but not in ridges.

Sandstone – is made of mineral grains from quartz, feldspars, gypsum, and rock grains from granite, basalt, limestone, etc. These grains are deposited by streams near highlands or waves and currents near a shore. The grains are compacted and cemented by silica, calcite, or iron oxide. Sandstone is resistant to erosion and weathering but produces a large amount of wasting.

Conglomerate Rock

Conglomerate – is sand with rounded pebbles that is naturally cemented together. It is deposited by streams near higher land and by waves near a shore. Conglomerate erodes like sandstone.

Breccia – is a mixture of angled rock fragments. Breccia is found near faults, near cliffs, and in submarine slides.

Metamorphic rocks – are a sedimentary or igneous rock that has been changed due to internal heat, penetration by fluids (hydrothermal solutions of mineral-rich steam), or pressure. These rocks form deep in the Earth's crust particularly in areas of mountain buildup. Sometimes, these changes can cause the atoms in metamorphic rock to rearrange and form different minerals. Sometimes, new chemical elements are rise up by hot steam traveling through the existing rock. When this happens, minerals are altered in the rock and new minerals are formed. They are eventually exposed to erosion. Most metamorphic rocks are hard and chemically resistant.

However, marble is not hard. Metamorphic rock textures are either foliated with cleavage or non-foliated with no cleavage. Metamorphic rock can be formed by contact with magma. The heat and pressure associated with the burial of sediment by pressure forces have an effect on its physical properties. Layered, or foliated metamorphic rocks form from the existing rocks that have many different mineral grains. For instance, shale, during metamorphism, contains small grains. Shale can reform into larger grains giving a layered appearance. Shale may also be altered into slate, schist, or gneiss, all of which contain foliation of various grain sizes.

Non-foliated metamorphic rocks form mono-minerals which are rocks composed of one mineral this is exposed to metamorphism. When limestone composed of calcite becomes marble, it is because of heat and pressure. Sandstone composed of quartz and feldspar will become quartzite.

Quartzite – is derived from a quartz sandstone. Quartzite displays a smooth, sugary appearance. It is dense and tough and is usually the most resistant rock in its locality.

Vishnu schist
1,745 million years old

TRAIL
OF
TIME

Schist – is derived from igneous rock or shale. It has interlocking crystalline grains of mica, quartz, and hornblende. It is arranged in sheets along where the rock breaks. Hardness increases with the amount of quartz content. Schist may be flakey and is resistant.

Slate – is derived from shale. It has interlocking, microscopic crystalline grains. It forms in sheets and breaks in that form. It is softer than steel and is resistant.

Gneiss – is derived from shale, sandy shale, or igneous rocks. It has visible interlocking crystalline grains of feldspar with some quartz with streaks or bands of mica.

Phyllite – is derived from shale. Similar to mica and schist in appearance, but is more lustrous, and silvery. Phyllite is flaky and its sheet forms may not be parallel. It is resistant.

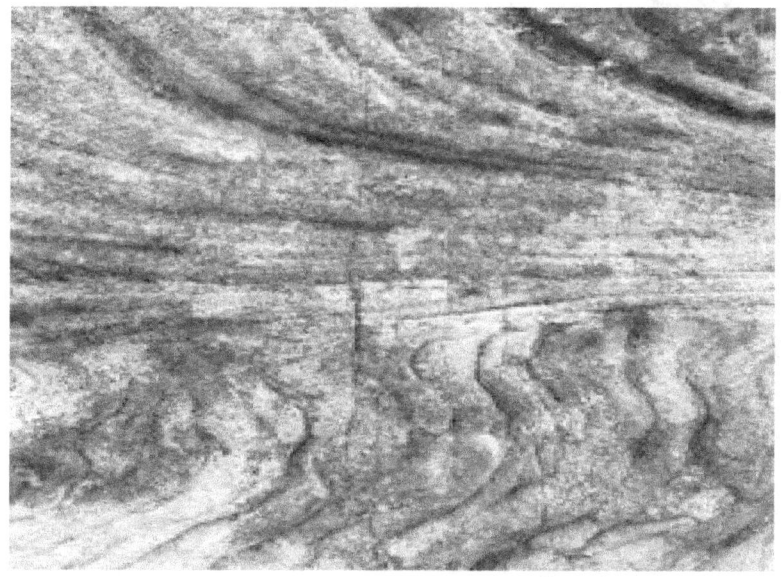

Marble

Marble – is derived from limestone. It has visible interlocking crystalline grains of calcite and dolomite. Marble is white when pure but may be streaked when combined with other minerals in its formation. Hardness is equal to a copper coin. Marble is found in temperate humid climates. It is found on peaks of small mountains.

The oldest known rocks found to date on the Earth's surface are igneous granites from northwest Canada. It is estimated to be 3.96 million years old. Some zircons have been dated at 4.3 billion years old. These are the oldest objects known on Earth.

GRANITE *GRANITO*	AGATE *AGATA*	LEPIDOLITE *LEPIDOLITA*	PETRIFIED WOOD MADERA *PETRIFICADA*
MAGNETITE *MAGNETITA*	GOLD ORE *MENA DE ORO*	OIL SHALE ROCA *ESQUISTOSA*	CALCITE *CALCITA*
SERPENTINE *SERPENTINA*	ROSE QUARTZ *CUARZO ROSADO*	COPPER ORE *MENA DE COBRE*	MICA *MICA*
LAVA *LAVA*	ALABASTER *ALABASTRO*	PYRITE *PIRITA*	FLUORITE *FLUORITA*
SANDSTONE PIEDRA *ARENISCA*	QUARTZITE *CUARCITA*	MARBLE *MÁRMOL*	LIMESTONE *PIEDRA CALIZA*
OBSIDIAN *OBSIDIANA*	SULFUR *AZUFRE*	FELDSPAR *FELDESPATO*	EPIDOTE *EPIDOTA*

Igneous, Sedimentary, and Metamorphic Rocks that are Common

Name	Description, Texture, and Mineral Composition
IGNEOUS ROCK	
Granite	Coarse-grained, quartz and feldspar
Granodiorite	Coarse-grained, quartz and feldspar
Diorite	Coarse-grained, feldspar, and abundant
Gabbro	Coarse-grained, dark-colored minerals abundant
Peridotite	Coarse-grained, dark-colored minerals, and olivine
Diabase	Medium-grained, dark gray minerals
Rhyolite	Fine-grained, light-colored minerals
Andesite	Fine-grained, dark-colored minerals
Basalt	Fine-grained, dark-colored minerals
Obsidian	Glassy; dense and dark-colored minerals
Pumice	Glassy; light-colored and lightweight minerals
SEDIMENTARY	
Conglomerate	Detrital; rounded, cemented particles
Breccia	Detrital; angular, cemented particles
Sandstone	Detrital; rounded, sand size particles
Siltstone	Detrital; silt size particles
Shale, claystone	Detrital; clay size particles
Limestone	Chemical; organic origin, calcite
Gypsum	Chemical; saline origin
Chert, flint	Chemical; siliceous origin
Coal	Chemical; organic origin
METAMORPHIC	
Slate	Foliated, often from shale, cleaves into plates
Schist	Foliated, elongated
Gneiss	Foliated, coarse-grained, often from granites
Quartzite	Non-foliated: quartz sand cemented with quartz
Marble	Non-foliated: from limestone or dolostone

Soil

Regolith is the blanket of loose, non-cemented, disintegrated rock particles and mineral grains formed from the weathering of bedrock. The soil is the part of the regolith that can support rooted plants. Chemical, physical, and biological factors control the soil formation process. Not all soils are alike. Soil types depend on the parent material and climate. Climate determines soil texture. Vegetation and topography which determines the thickness of the soil layer usually is from one to six feet (30 to 200 centimeters). Plus the length of time in the soil making process all are factors that determine soil. The soil classification devised in 1975, is the most widely used system around the world. Soil is classified on the observable profile soil characteristics. Each group or order, is divided into sub-orders, then into great-groups, subgroups, families, and series. There are 10,000 different kinds of soil at the series level.

Soil Order	Description
Alfisols	Gray-brown surface horizons, subsurface clays
Aridisols	Dry for more than six months per year; low organic matter
Mollisola	Black, organic surface horizons
Spodosols	Amorphous materials in subsurface horizons
Ultisols	Moist soils with clays
Histosols	Organic Soils
Inceptisol	Moist soil, generally influenced by parent matter
Entisols	No pedogenic horizons

Crystals

Most minerals can form crystals under the right conditions. When the atomic structure of any material is arranged in an orderly manner, a crystal is formed. Crystals are described by their 3-dimensional shape formed in atoms which occur in one of seven crystal system shapes. They form in a polyhedron which is a solid whose faces are polygons.

Hexagonal system – Hexagonal crystals are composed of two axes aligned to each other at 120-degree angles to each other. The third axis is at right angles to the two others at a different length.

Pyrite with quartz attached

Isometric (Cubic) systems – Isometric crystals have three axes of equal length at right angles to each other. Diamond, fluorite, galena, pyrite, halite, and garnet crystals are isometric.

Monoclinic system – Monoclinic crystals have two of the three axes not at right angles but the third axis at right angles to both and none are equivalent. Azurite, muscovite, malachite, gypsum, and borax crystals are classified as monoclinic.

Orthorhombic system – Orthorhombic crystals have three axes at right angles to each other with none equivalent. Chalcocite, topaz, sulfur, marcasite, and stibnite crystals are orthorhombic.

Tetragonal system – Tetragonal crystals are composed of rectangles, with three axes at right angles to each other and two or three of the axes equivalent. Zircon, rutile, and cassiterite crystals are tetragonal.

Triclinic system – Triclinic crystals have no axes at right angles and none equivalent. These crystals are rare. Sanidine and turquoise crystals are triclinic.

Trigonal systems – Trigonal crystals are composed of one vertical threefold axis that develops similarly to the hexagonal system. Dolomite, corundum, cinnabar, and arsenic crystals are trigonal.

Fake diamonds – Scientist have been trying to create the hardest naturally occurring mineral known on Earth, diamonds.

One failed attempt involved melting carbon at the center of an iron ball. The hope was that if the carbon was heated and then rapidly cooled, a diamond would be produced. The experiment failed. Another experiment was to use explosives, hoping that the heat and pressure of the blast would create a diamond. Instead, the carbon turned to graphite.

Tiny industrial diamonds called 'grit' have been made by heating certain carbon compounds to 4,900 degrees F. (2,704 degrees C.) at a pressure of 1.5 million pounds per square inch (105,000 kilograms per square centimeter). Nature makes the largest and best diamonds.

Interesting crystals – It's important to note that the arrangement of atoms determines the characteristics of the mineral. For instance, a diamond, which is the hardest substance known, and graphite, which is used as a lubricant, are both made entirely of carbon atoms. The diamond is one huge isometric carbon molecule. Graphite has the atomic formation in layers of flat hexagons which slide relative to each other. Scientists have discovered another interesting carbon chemical called C-60 where the molecule consist of 60 carbon atoms. It is nicknamed a bucky-ball and resembles a soccer ball with unusual magnetic, chemical, and electric properties.

Salt, (Sodium Chloride)

Diamonds – are extremely rare. In the 1800's, diamonds were discovered in alluvial deposits. Because of their size and density, the loose diamonds were found along primitive shores. Most diamonds are found in the Earth's mantle layer. Some areas in the Earth's crust that had experienced high pressure metamorphism have yielded diamonds. In these areas, the diamonds have not been transported by any natural means. When a good sized meteorite strikes the ground, the pressure and heat generated can produce micro-diamonds. This type of meteorite impact occurred in Russian which has the world's largest deposit of this type.

Diamonds are not formed from coal. Coal is derived from plant based materials. Diamonds were formed before plants formed on Earth. Diamonds are typically found in the oldest parts of the Earth where the cores are more stable. However, an exception is Australia which is the largest producer of diamonds in the world. This area has undergone compressional tectonics. The diamonds here are found in lamproite rock whereas diamonds are usually found in kimberlite rock.

Kimberlites are found in narrow sills and dikes that are about 1 to 4 meters and in large pipes that can be from 75 meters to 1.5 kilometers. The physical feature of kimberlites is that the color is dark blue-green to a gray-green. However, after exposure it turns brown and crumbles. These rocks are a mixture of minerals that rise up from the mantle. The three diamond bearing rocks, kimberlite, lamproite, and lamprophre, lack certain minerals with the formation of a diamond.

Kimberlite pipes are difficult to locate. They have lower topographic relief locations than the surrounding rock. They also weather shortly after they are exposed, usually within a few years. Kimberlites cover the diamonds with lakes, soil, sediment, or vegetation. Modern methods in searching for diamonds are geophysical. The methods include, electrical resistivity, aeromagnetic surveys, and gravimetry. Geologist also use tools such as isotropic dating to identify the region likely to contain diamonds. The geologist then collect the samples to identify kimberlite fragments. The best approach is geothermobarometry. This method permits the geologist to analyze the minerals in the sample in equilibrium with mantle materials. Existing mines have a lifetime of 25 years. As rare as diamonds are, a shortage of diamonds may occur in the near future.

EXPERIMENT:

- Obtain an empty egg carton.
- Locate various rocks.
- Perform a series of test on each rock.
- Record your findings.
- Sort the rocks in your rock collection display.
- Label each rock with its geological name.

Rock and Mineral Collecting

Minerals include a variety of metallic ores and non-metallic ores. Metals are important throughout the entire world. Metallic ores are needed to produce iron and steel. The non-metallic ores include a variety of uses such as building materials, fertilizer, and much more. Other minerals and rocks are gemstones used in jewelry. Ancient civilizations searched for deposits of obsidian and flint for arrowheads. Explorers were in search of gold. When you collect rocks and minerals, you are obtaining Earth's process of geological history.

Listing of rocks and minerals and where to find them.		
Rock or Mineral	Common Name	Where to Obtain Them.
Limestone	Road aggregate or chat	Building supply store
Volcanic scoria or pumice	Lava rock	Landscaping supplies
Marble	Marble chips	Landscaping supplies

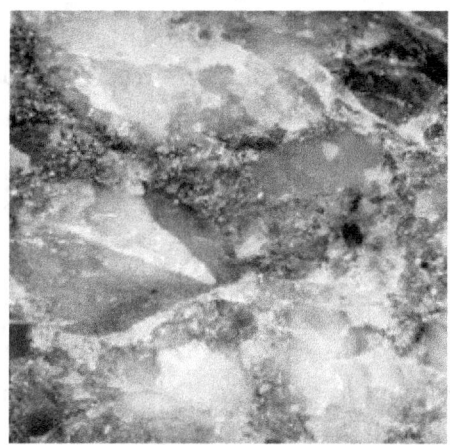

Gold ore

Gold – is very rare. It is estimated that 187,000 metric tons of gold exist above ground. The oceans contain more gold than is on land. The amount of gold above land, if formed together as a cube, each edge would be 21 meters. That amount of gold would be worth more than $8.9 trillion dollars with the troy ounce of $1,349. In recent times, China was the number one gold producer at 455 metric tons. Australia was second at 270 metric tons, followed by Russia, at 250 metric tons.

During the 1800's, gold rushes occurred at any location where gold was discovered. The first documented gold rush occurred in the United States in 1803 in North Carolina. Further gold rushes occurred in California, Colorado, the Black Hills, Otago in New Zealand, Australia, South Africa, and in the Klondike area of Canada.

Although one-quarter of the world's gold mining is with small scale mine operations, the majority of gold mining is achieved with large scale mining operations. These operations function economically in that they embark on easily mined gold deposits. Recently, the average gold mining and extraction cost about $317 per troy ounce.

Gold is refined by the electrolysis process known as the Wohlwill process. Another gold refining process is called the Miller process that uses chlorination in the melt.

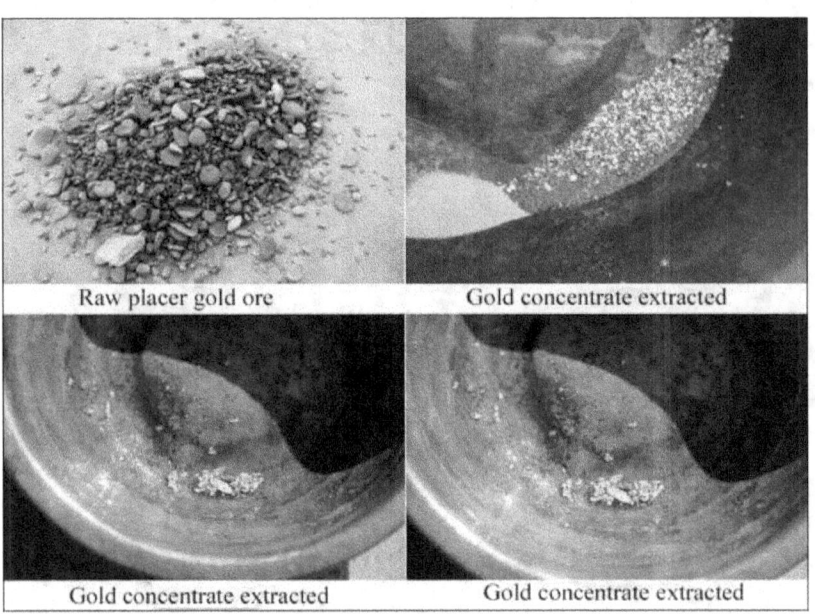

Raw placer gold ore | Gold concentrate extracted

Gold concentrate extracted | Gold concentrate extracted

Normal Uses of Rocks

Some geologists find new uses for the rocks and minerals found. Most cities have a nearby source of earth materials used for manufacturing, energy production, construction, agriculture, and recreation. The construction of buildings and roads depend on the sand and gravel operation nearby. The most commonly mined and processed mineral is crushed limestone. Millions of tons of crushed limestone are mined each year for manufacturing of cement, or crushed rock called aggregate. Aggregate is used for road concrete, highways, and driveways. Other common mining operations include coal, ore, gypsum, salt, fertilizer production, and oil production. Almost everything around you is a product of a mining operation. If you look around, you will find examples such as your home, school, church, sports field, camp, tools, and the list goes on.

Building roads – is certainly a way that geologists impact our lives. Civil engineers and municipalities employ geologist plan to build the best highways, bypasses, exits, and streets. The locations for this type of development is examined by these professionals for the best possible sites. Cost, the type of material needed, transport of materials, are just a few factors that are taken into planning.

Activity 8   Road Construction Materials in Your Community

Understanding the types of rocks that make up Earth, research how roads are constructed in your community. In this activity, you will identify the three most common road building materials in your community, how they are produced, and how they are used in road construction.

Strategy:

Step 1 With the help of a responsible person, contact a construction engineer or concrete or construction business listed in your phone directory.

Step 2 Record notes as you ask what the business uses as its three most common road construction materials.

Step 3 Ask how these three materials are produced and how these three materials are used in road construction.

The term 'fossil' is derived from the Latin 'fossilis' which means "something dug up". Fossils are the remains of ancient organisms preserved close to their original shape. The standard method used to divide the Earth's natural history into parts is the geologic time scale. The units of time on the scale are long. The life of organisms and geological events are recorded, and the information is revealed by the fossils left behind. Scientist use radioactive dating of rocks and fossils and absolute dating to analyst Earth's history. Relative dating is the basic technique. The geologist observes and measures the layers and distance from the surface to determine the age of rocks and fossils. It is the layers that are measured and compared to other layers, above and below it that is analyzed. The layers age is determined by the other layers in a relative position to each other.

Scientists then use absolute dating to compare and confirm the relative dating. Absolute dating is the measurement of radioactive decay. The atomic structure diminishes over time as they change from one form of an isotope to another form. For instance, potassium becomes argon and uranium eventually becomes lead. Scientists use the time needed to decay. This event is called 'half-life'.

Half-Life – Radioactive decay is when the nucleus of an atom loses some of its atom particles over a period of time. When the atomic particles leave the nucleus, the atomic number changes and the element convert to another product, sometimes a completely different element.

The four radioactive isotopes that occur in rocks are two isotopes of uranium (U238) and (U235), rubidium (Rb87), and potassium (K40). Geologists use the ratio difference to determine the rock's half-life. A 1 gram sample of uranium 238 would take 4.5 billion years for half of its life to decay to 0.5 grams and to decay to lead 206. The age of the Earth is believed to be 4 billion, 600 million (4,600,000,000) years old.

Radioactive Atomic Half-Lives	
Atom	Billion Years
Uranium 238	4.5
Uranium 235	0.7
Rubidium 87	47.0
Potassium 40	1.3

Paleontology

Paleontology is the study of ancient life. The geologist that study fossils are called paleontologist. Clues to the Earth's past environment, climate, and living organisms are determined by the preserved remains of plants and animals decay before they were fossilized. Fossil finds are rare and only represent the billions of organisms, plant and animal, that have lived on Earth. If an animal dies in the forest it won't turn into a fossil because other animals, including insects, would consume its carcass. The conditions must be right for a fossil to form. The animal must be buried soon after death. Animals and plants that die on land have almost no chance to become a fossil. However, if the plant or animal falls into a body of water or mud, the soft parts of the animal or plant may still decay, but the firm parts of the organism, such as bones or plant veins, will be protected. These are often replaced by mineralization, where the original living matter is replaced by minerals. Other fossils include molds, cast, or imprints of dinosaur footprints. Dinosaur National Monument in Utah, and the La Brea Tar Pits in Los Angeles have evidence of the prehistoric animals that have fallen in these areas. Occasionally, when shells, or other organisms have fallen in soft mud, the shape of the shell or organism is preserved as a mold as the mud turns to rock. The largest fossils are not the oldest. The microscopic, single-celled organisms lived 3.75 billion years ago. The larger organisms, including trilobites, lived in the oceans 600 million years ago. The first amphibians arrived about 380 million years ago, the reptiles appeared 320 million years ago, and dinosaurs emerged 225 million years ago.

Paleontologist estimate that the first dinosaurs evolved on Earth 225 million years ago and they became extinct 65 million years ago.

- Dinosaurs are members of a group of reptiles known as archosaurs (ruling reptiles). They are divided into two main orders, the Saurischia (reptile hipped) and Ornithischia (bird-hipped).

- All dinosaurs were land-living creatures. The gigantic sea creatures of the Mesozoic era were not dinosaurs.
- Pterosaurs were not dinosaurs. They were flying reptiles that looked like lizards with wings.
- Today's birds and crocodiles are thought to be the closest relatives to the dinosaurs. All creatures on Earth are descendants of creatures that lived during the dinosaur era.
- The Brontosaurus's correct title is Apatosaurus. When paleontologists were classifying the dinosaurs at the turn of the last century, the wrong head was placed on the Apatosaurus. Brontosaurs never existed.

Questions about dinosaurs have not yet been proven. Were they warm-blooded or cold-blooded animals? Were they born cold-blooded and then became warm-blooded, or vice versa? Did the dinosaurs become extinct rapidly or slowly? Was their extinction due to disease, or a meteorite, or climate change from a too much volcanic activity?

Past Extinction

Over the past 600 million years there have been numerous extinctions of plant and animal species. Records of fossils in rock layers have revealed that there have been six major extinctions over time. One major extinction occurred 250 million years ago during the Permian period when 96 percent of marine species died. More famous is the extinction of the dinosaurs which occurred 65 million years ago. The majority opinion is that objects from space struck the Earth, creating dust and debris to enter the Earth's atmosphere that blocked sunlight. This caused tsunamis, earthquakes, and climate change and altered the food chain.

Geologists refer to the abundant fossil record as the time when numerous and different life forms roamed the Earth at the same time. Before the Cambrian Period, the life forms were single-celled organisms. This included bacteria and plankton. During the beginning of the Cambrian era, multi-celled organisms populated the Earth. Some organisms developed bodily functions that didn't depend on photosynthesis. These organisms were growing in size and population. When a set of conditions, as evidenced by fossils change and the plants and animals are no longer able to obtain food and dies off, the change is considered a paleoenvironmental change. During these events, individual species and entire ecosystems became extinct.

EXPERIMENT

Make a fossil using plaster of Paris and a disposable aluminum baking tin.

- Obtaining proper permissions, select an area to prepare your fossil production.
- Decide whether to create a fossil of an animal, a human foot, a leaf or all three.
- If you are making a fossil of an animal track, cut a strip of thick paper and fasten both ends together. Surround the track with your circular paper form.
- Mix the plaster so that it is fairly thick, yet will penetrate the minute details of the track.
- If you decide to create a human footprint fossil, rum petroleum jelly all over the foot. Pour the plaster into the aluminum baking pan to fill one inch or so from the bottom. Make sure that the plaster is thick. Step into the pan but not too deep. The foot must not touch the bottom. Wait fifteen minutes or so, then remove the foot.
- Follow the same instruction above for the leaf fossil.
- As with all experiments, clean the area completely as soon as you are done.

Not all species of dinosaurs lived at the same time. The Apatosaurus (formerly known as the Brontosaurus) did not roam the Earth at the same time that the Tyrannosaurus rex lived. The final era is the Cenozoic Era which began at the end of the Mesozoic Period and continues till today. During this time the last ice had occurred. This era produced the development of glaciers in the Northern Hemisphere. Ice measured more than a mile thick in some areas. Many animals adapted to the freezing temperatures. Such species of animal are the mammoths and mastodons. Many were hunted for food. The climate south of the ice sheets was rainy and swarms of insects inhabited that area. The bats and birds of that time, that ate the insects, were probably much larger than the bats and birds of today. Every plant and animal species are found within the habitat in which they are best adapted for survival. The environment varies from one habitat to another. Smaller plants grow thickly as the carpet the floor of the rainforest. The same plants would not survive in a field or any hot, dry areas.

Because animals tend to live in very specific environments, their fossil remains give geologist information about the history of the area. Geologists study fossils of marine animals and their environment. The knowledge of this information assists geologists locate areas of possible mineral value.

Field trip – Visit a museum or university to view their rock and fossil collection. Find out how geologists at that location organize and categorize their collections.

- Photograph or sketch fossils that you have found on display. Include a written description of the fossil.
- Note the type of rock that contains the fossil. It will most likely either be limestone or sandstone. Are the rock grains fine or coarse? Finer grains maintain greater detail. What type of environment did it come from? The fine grains come from a calm flow area, while the coarse-grained rock comes from an area with a current.

Natural Wonders of the World

- Grand Canyon, United States
- Iguazu Falls, Argentina
- Rio de Janeiro Harbor, Brazil
- Niagara Falls, United States
- Natural Bridge, United States
- Yosemite Valley and the Giant Sequoias, United States
- The Nile River, Egypt
- Mt. Everest, Tibet and Nepal
- Matterhorn, Switzerland, and Italy

Glaciers continuously shape landforms. They grind lowlands, carve mountain peaks, and cut out valleys. Glaciers also spread rock waste, fill lake basins, deplete the oceans, and depress the Earth's crust. Glaciers cover a tenth of all land, particularly on mountain tops and within the Polar Regions.

Glaciers form where the annual snowfall is above the annual melting and evaporation of snow. When the snow accumulates above 100 feet or more, it compacts the snow at the bottom tightly. The snow at the bottom melts slightly and then refreezes as the temperature fluctuates. At that point, it takes on a grainy form called 'neve' and slowly turns into solid ice. Then, with the overlying pressure and weight, begins to spread like mud would spread. As more snow accumulates, the ice moves out from the center as a sheet or stream. This is now called a glacier.

Valley glaciers originate in the mountain hollows and move downward through the valleys previously cut by streams. The speed of the movement is measured in inches per day. The movement is caused by the melting and refreezing within the glacier and by slippage of the ice crystals. When the ice goes over large humps, the glacier tends to spread out against the valley walls or it widens where the valley widens. Tensions develop because of the spreading, thus causing fractures called 'crevasses' to form. Weathered rock that falls on to the glacier becomes a moraine. When the ice stream is joined by tributary glaciers from side valleys, another moraine band is formed.

In temperate zones, the ice of a valley glacier descends until it reaches a level where the ice melts thus, the glacier ceases to exist and a stream forms. As the ice melts, it may have accumulated a large amount of rock waste forming a crumbling wall with an ice arch. This is the opening of a subglacial tunnel from which meltwater flows. This water mixes with the ground rock and generally yields a turquoise hue to many glacial lakes. If the ice front is stationary because the rate of ice arrival exceeds that melt rate, a terminal moraine accumulates at that point. In polar lands, glaciers enter the ocean forming icebergs.

Hanging glaciers – are ice masses that grow in mountain hollows and terminate at the top of a steep slope.

Piedmont glaciers – are broad ice masses form on lowlands as valley glaciers spread.

Ice caps – are thick ice accumulations that cover a group of highland summits.

Ice sheets – are large ice masses that spread radially from the highland centers of the accumulated ice and cover the lowlands to the depths of thousands of feet.

Today, glaciers cover 7 to 8 million cubic miles of the Earth's 325 cubic miles of water. Ice sheets cover most of Greenland to a maximum of 10,000 feet in depth. All of the Antarctica glaciers cover 14,000 feet in depth. The weight has depressed the central region of Antarctica to a depth of a half mile. Piedmont glaciers are restricted to the Antarctic, Greenland, and the Alaskan coast. Large valley glaciers are found in the southern Canadian Rocks, Alps, Himalayas, and at the northern Andes. Small glaciers can be found in temperate areas and tropics such as Kilimanjaro.

The most recent ice age was during the Pleistocene Epoch. It began one and a half million years ago as air temperatures dropped 14 degrees F. below the present average. Ice domes in the north spread to middle latitudes, which covered one-third of the land. Some of the lowlands were under more than 2 miles of ice. Ice loaded terrains sank 1,000 feet. The sea level was down 200 to 300 feet, exposing the continental shelf areas. As the temperature rose, melting was accelerated. Rock waste from the glacial erosion was deposited, basins were filled with meltwater, and the sea level rose. Scientists have found that there have been four glaciations (ice ages). Each period lasted between 50,000 to 100,000 years. The fourth ice age climaxed 18,000 years ago. The causes for these ice ages were due to massive volcanic dust in the atmosphere, reduced solar radiation, and shifts of the planets axis of rotation.

Glaciers cause erosion. They shape the land as they pass over it. Rock fragments in the ice rub against the bedrock and gouge it. Where the glacier is attached to the bedrock and then moves, it plucks out parts of the bedrock.

Terrains of low or moderate relief — This is the terrain that is sculptured by ice sheets. Areas that have been plucked out form basins for ponds.

Mountains in an ice sheet — are mountains that are smooth when overridden by an ice sheet when the ice was at its thickest. When the ice starts to melt, it widens the valley.

Mountains outside of an ice sheet — become shaped by the valley glaciers. These mountains have sharp peaks and enlarged valleys.

Cirques — are hollows at the valley heads where the valley glaciers originated. Well-developed cirques have semicircular cliff with a basin at the foot. When the ice melts, the basin may hold a pond.

Aretes — are narrow, jagged ridges between cirques that have formed as headwall cliffs erode back.

Cols – are low saddles between cirques. They are formed by the intersection of curving cirque walls. They are often referred to as mountain passes.

Horns – are peaks made by the retreat of cirque headwalls from several sides.

Fiords – are glacial troughs with floors now below sea level.

Truncated Spurs – are snubbed off ends of ridges along sides of a valley.

U-Shaped troughs – are valleys shaped by moving glaciers into troughs which adjust to the flow. When the ice melts, a trough with a U-shaped cross-section appears.

Rock basins – are hollows made by glacial erosion in weaker rock. Many have ponds.

Hanging valleys – are tributary valleys of low gradient that quickly terminate in a steep wall of a large trough. Often, these sites have waterfalls.

Rock steps or stairs – are found along the trough floor, mainly near the upper end of the main trough at junctions with tributary troughs or at sites of former icefalls.

Rock drumlins – are like rock knobs, but with both front and rear ends smooth.

Striae or striations – are grooves made by rock fragments embedded in the ice. They can be more than a foot deep and several feet wide.

Crescentic fractures – are curved breaks, usually in nested series that were made by embedded boulders bumping over bedrock.

Crescentic gouges – are depressions made by the removal of rock chips by abrasive boulders.

Glacial Deposits

Glacial erosion causes much rock waste. Waste from the Pleistocene valley glaciers was left in mountain valleys. Most are washed by meltwater out on to lowlands. Deposits from ice sheets may be found at various levels, including mountain summits. Glacial rock waste is called 'drift' by geologists. Drift is sorted and stratified when it is deposited by water. Drift is unstratified and unsorted if it is deposited by ice action. Unsorted drift is called 'till'.

Terminal moraines – are till accumulations pushed up when the glacier front is almost stationary.

Lateral moraines – are long, low ridges of mass-wasted rock carried along the glacier's sides.

Eskers – are winding ridges of stratified sediments deposited by streams that ran on or within or below a glacier.

Kames – are irregular, rounded cone or dome-shaped hillocks of stratified drift that is deposited by meltwater running off glacier sides into melt holes.

Kettles – are depressions in drift due to the melting of buried ice blocks. They usually have ponds.

Drumlins – is till molded by the ice sheet into streamlined hills that are elongated toward the direction of the ice movement. The drumlins can be 500 feet high and several miles long.

Till plains – are deposits that completely bury pre-glacial landscape. It consists of ground moraine, the material deposited directly by the glacier melting.

Erratics – are rock fragments that is transported by a glacier to a distant place and deposited as isolated boulders, Erratics are prominent on smooth surfaces. Their appearance is often different from the bedrock where they are found.

Valley trains – are deposits of sorted drift extending down the valley from moraine in a glacier trough.

Outwash plains – is a sorted drift spread widely as a low alluvial fan that is fringing on the end moraine of a continental ice cap.

Mountains are the result of an intense deformation of the Earth's crustal belts. Some mountains are the direct result of faulting and folding. Other mountains are due to a buildup of volcanic eruptions that were caused by the deformation of the crust. Some mountains are very high while other mountains appear to be low but on their way up. A number of mountains are slowly eroding over time.

Orogeny – is the development of mountains. These mountains begin with the accumulation of sediments in a large, ocean-filled, down-warp called a 'geosyncline'. This can be a 1,000 miles long and more than 300 miles wide. The sediment may be 8 miles thick. Eventually, the sediments are melted under great pressure. At this point metamorphism and igneous activity occurs within the geosyncline and along its edges. Overlying sediments are folded and thrust faulted by the pressure of the compression. The large mass rises and faults form. This is the beginning of mountains.

Crustal activity in mountains continues over long periods of time. When mountains are eroded, others may form and rise near them. Former mountains become geosynclines. Igneous and metamorphic rocks are the bases of the earlier mountains. One cause of uplifts is erosion. When a mountain block's top erodes, the block rises. The mountain block achieves the same level of density as the surrounding blocks.

Mountain types – There are four main categories of mountains. They are foreland-fold, complex, volcanic, and fault-block mountains.

Foreland-fold Mountains – consist of regular folds with thrust and tear faults. The faults increase and displacement occurs toward the geosyncline center.

Complex mountains – form near a geosyncline's center. Faults and folds become more complex. Recumbent folds with a large over-thrust slice occur. Rocks become metamorphosed and are injected with magma. They are bordered by fore-land folds or plateaus where sediment is stripped, exposing an older, worn complex mountain mass.

Bold escarpments – occur were over-thrusting or recumbent folds have brought together rocks of unequal resistance to erosion.

Plateau-like masses – are underlain by low-dipping beds. They have an over-thrust on one side of the recumbent fold that is partially eroded.

Long ridges – are curving back on to themselves with a gently plunging large fold.

Clusters or Irregular hills – are igneous intrusive masses. Some might be high due to high resistance.

Fault-block Mountains – are a result of gravity faults. It starts when a broad uplift stretches the crust. When the crust's tension is compromised, fractures occur and the weight of the crustal blocks causes a collapse. Faulting then occurs with tilt blocks, grabens, and horsts as a result. Block Mountains occur where earlier deformations had occurred such as a basin or range area. Numerous blocks directed in a north-south direction lie between superblocks. This occurs when the interior blocks, sunk in relation to the super-blocks which form a super-graben. Block Mountains may appear in areas of extensive basaltic lava flows. On ocean bottoms, gigantic block mountains parallel the volcanic active rift zones.

Volcanic Mountains – build up by long, continued activity around volcanic vents. These ranges rise up from the seafloor. Continental volcanoes produce felsitic and andesitic lavas caused by melting and re-melting of continental crust during the mountain building. In continents and on the sea bottom, rifts from basaltic lava erupts. The eruption may be arched into a mountain form. It is then surmounted by cones of a felsitic composition. Volcanic mountains endure long after the volcano has become extinct.

Major Mountain Peaks

Name	Mountain Range	Location	Height Above Sea Level (feet)(m)
Everest	Himalayas	Asia	29,022 8,846
K2 (Godwin Austen)	Himalayas	Asia	28,250 8,611
Kanchenjunga	Himalayas	Asia	28,208 8,598
Lhotse I	Himalayas	Asia	27,923 8,511
Makalu	Himalayas	Asia	27,824 8,481
Lhotse II	Himalayas	Asia	27,560 8,400
Dhaulagiri	Himalayas	Asia	26,810 8,172
Manaslu I	Himalayas	Asia	26,760 8,156
Cho Oyu	Himalayas	Asia	26,750 8,153
Nanga Parbat	Himalayas	Asia	26,660 8,126

Highest and Lowest Continental Locations		
Name	Continent	Height or Depth (feet) (M)
HIGHEST Points Above Sea Level		
Mount Everest	Asia	29,022 8,846
Mount Aconcagua	South America	22,834 6,960
Mount McKinley	North America	20,320 6,194
Mount Kibo (Kilimanjaro)	Africa	19,340 5,895
Mount El Brus	Europe	18,510 5,642
Vinson Massif	Antarctica	16,860 5,139
Mount Kosciusko	Australia	7,310 2,228

The Dead Sea

LOWEST Points Below Sea Level		
Lake Eyre	Australia	52 16
Caspian Sea	Europe	92 28
Salina Grandes	South America	131 40
Death Valley	North America	282 86
Lake Assal	Africa	512 156
Dead Sea	Asia	1,299 396

Since Earth's earliest days, igneous activity has periodically occurred. The heat causes the magma to rise. The lava travels through crustal fractures. A magma chamber holds the hot lava. Weaker areas of rock above the magma chamber crumbles and allow the magma to rise. An eruption occurs when the surface can no longer hold back the pressure. Areas of high stress usually results in volcanic activity particularly in the concentrated stress areas, along young mountain areas and rifts. The areas of volcanic activity has changed from age to age. The piling up of lava from sea vents probably caused the first areas of land. Even today, this process is creating islands. Many of the islands have evolved from igneous activity and rock waste from erosion of old volcanic highlands. More recent volcanic activity results in cone-shaped landforms. Many great mountain ranges were a result of volcanic activity. Volcanic land forms are a result of the volcanic activity. Eruptions may be slow and thus creating an expansion of the surface, or they may be explosive where much of the material is blown away from the igneous site.

Lava flowing down a street in Hawaii in 2018

There are four types of volcanoes:

1.   Shield volcanoes are formed by layers of lava flows.

2.   Cinder volcanoes are made up of small lava fragments. They form slopes from 30 to 40 degrees.

3.   Composite volcanoes are created with alternating layers of ash and lava.

4.   Lava domes are very thick layers of lava.

Volcanic cones have unique features. The features are spines, plug domes, spatter cones, parasitic cones, and pit craters.

Spines – renewed activity produces towering masses of solid lava which may rise hundreds of feet. The internal weight and pressure would cause the cone to burst open. The hardened lava flakes off adding to the base.

Devil's Tower a volcanic core

Plug domes – Where masses of lava well up from a vent or crater and solidifies into a rounded form is called a plug dome. This phenomenon may occur over a hidden vent, or in a crater, or may flank the primary cone.

Spatter cone – These cones are formed when lava is spattered from a vent or near the primary cone.

Parasitic cone – These cones build up on the sides of the primary cone by the emission of lava through the vents.

Pit craters – When the lava is withdrawn at a vent, a depression occurs forming a pit crater.

Calderas are craters that are enlarged by the collapse of walls due to undermining. The floor exhibits small domes produced by pressure below the surface. A new cone may form after a period of dormancy as a result of newer eruptions.

There are two types of material resulting from an eruption. Pyroclastic material is blown out of volcanoes as fragments. They are of various sizes and shapes. The other type of material is the lava flow coming from pipes and fissures.

Lava flow toward the sea in Hawaii 2018

Pyroclastic material includes volcanic ash, volcanic tuff, volcanic dust, lapilli, tuff breccia, cinders, scoria, volcanic blocks and volcanic bombs.

Volcanic ash are particles that are 1/100 to 1/7 inches in diameter.

Volcanic tuff is compacted volcanic ash or dust. Occasionally this volcanic tuff is formed by the heat of the volcano. Streams often dissect the tuff forming cliffs or pillars.

Volcanic dust are particles that are smaller than volcanic ash.

Lapilli are tiny stones 1/7 to 1 inch in diameter.

Tuff breccia is a mixture of angular lava fragments with volcanic tuff.

Cinders are lava fragments from 1 to 3 inches in diameter. The may be crystalline and glassy and they contain several gas holes.

Scoria is heavier the pumice. It is basaltic with several holes that formed by escaping gas. The scoria lava fragments are usually very dark in color.

Volcanic blocks are fragments of rock torn from the walls of the volcanic pipe during eruptions.

The solid form of lava rock is basalt. The basaltic lava flows from fissures and cones spread out and form plateaus and plains. Lava may form many shapes. Pillow lava forms underwater. Pahoehoe lava forms rope like coils. Lava stalactites form from dripping lava; usually in a cavern or tube. Pressure ridges form from the pressure below the surface. Basalt columns forms when compact lava is jointed and then cools and contracts. These columns are often six-sided. Squeeze-ups are formed when lava is forced up through small breaks in the crust. They form a bulb or mushroom shape.

Felsitic lava flows are fluid in movement. It may plug a vent and form a plug dome. If gas is below the plug, it may burst. The felsitic lava flows cool and solidifies forming light-colored felsite and rhyolite. The frothy lava solidifies to form pumice which is light enough to float on water. The felsitic lava that lacks gas cools to form a non-crystalline glassy rock, black in color, called obsidian.

In arctic zones and temperate zones, erosion of lava is resisted. The highly permeable pyroclastic lava in the form of cinder cones absorb water and experience little stream erosion. However, they are susceptible to wave action and erosion does occur. Highly permeable cinder cones absorb water and experiences little stream erosion. Basaltic lavas in the South Pacific have eroded because of the warm, humid tropical weather. The less permeable lava will gully quickly. During eruptions, heavy rains condense quickly into a stream that cuts deeply and directly from the cone. Hot mud pours down the gullies and settles on the lower slopes.

Volcanoes that have become extinct, or dormant, retain residual heat in the crust. This produces other geological phenomena such as hot springs, mud pots, and geysers. Groundwater that has percolated through a hot zone through tubes and joints, will rise to the surface forming a pool. These hot springs may have water several thousand feet below the surface. The hot springs that are full of minerals keep churning mud and the bubbling gas and mud rise to the surface to form mud pots. Geysers are hot springs that spew hot water and steam, then stop. Cooler groundwater enters a hot spot and is heated. The escaping water and steam pass through the vent until that amount of water has been processed. The geyser stops until the next volume of cool water becomes heated and rises to the surface.

Active Volcanoes:

Africa and Indian Ocean

Name	Location	Year
Lengai Ol Doinyo	Tanzania	1993
Nyamuragira	Zaire	1992
Piton de la Fournaise	Zaire	1992

Antarctica

Name	Location	Year
Mount Erebus	Ross Island	1990
Big Ben	Heard Island	1986
Deception Island	South Shetland Island	1970

Asia

Name	Location	Year
Aso	Japan	1993
Kralcatau	Indonesia	1993
Mayon	Philippines	1993
Sakura-jima	Japan	1993
Sheveluch	Russia	1993

Central America and the Caribbean

Name	Location	Year
Arenal	Costa Rica	1994
Pacaya	Guatemala	1994
Santiaguito Dome	Guatemala	1993
Ricon de la Vieja	Costa Rica	1992
Poas	Costa Rica	1992

Europe and the Atlantic Ocean

Name	Location	Year
Stromboli	Italy	1994
Ema	Italy	1993
Hekla	Iceland	1991

North America

Name	Location	Year
Cleveland	Alaska	1994
Kanaga	Alaska	1994
Kilauea	Hawaii	1993
Seguam	Alaska	1993
Akutan	Alaska	1992
Spurr	Alaska	1992
Westdahl	Alaska	1992
Colima	Mexico	1991
Mount St. Helens	Washington	1991

Australia, New Zealand, and the Pacific Islands

Name	Location	Year
Mount Semeru	Indonesia	1994
Langila	Papua New Guinea	1993
Ulawun	Papua New Guinea	1993
Manam	Papua New Guinea	1992
Ruapehu	New Zealand	1992
White Island	New Zealand	1992

South America

Name	Location	Year
Galeras	Columbia	1993
Guagua Pichincha	Ecuador	1993
Copahue	Argentina and Chile	1992
Lascar	Chile	1992

The escaping gas. The scoria lava fragments are usually very dark in color.

Volcanic blocks are fragments of rock torn from the walls of the volcanic pipe during eruptions.

The solid form of lava rock is basalt. The basaltic lava flows from fissures and cones spread out and form plateaus and plains. Lava may form many shapes. Pillow lava forms underwater. Pahoehoe lava forms rope like coils. Lava stalactites form from dripping lava; usually in a cavern or tube. Pressure ridges form from the pressure below the surface. Basalt columns forms when compact lava is jointed and then cools and contracts. These columns are often six-sided. Squeeze-ups are formed when lava is forced up through small breaks in the crust. They form a bulb or mushroom shape.

Felsitic lava flows are fluid in movement. It may plug a vent and form a plug dome. If gas is below the plug, it may burst. The felsitic lava flows cool and solidifies forming light-colored felsite and rhyolite. The frothy lava solidifies to form pumice which is light enough to float on water. The felsitic lava that lacks gas cools to form a non-crystalline glassy rock, black in color, called obsidian.

In arctic zones and temperate zones, erosion of lava is resisted. The highly permeable pyroclastic lava in the form of cinder cones absorb water and experience little stream erosion. However, they are susceptible to wave action and erosion does occur. Highly permeable cinder cones absorb water and experience little stream erosion. Basaltic lavas in the South Pacific have eroded because of the warm, humid tropical weather. The less permeable lava will gully quickly. During eruptions, heavy rains condense quickly into a stream that cuts deeply and directly from the cone. Hot mud pours down the gullies and settles on the lower slopes.

Volcanoes that have become extinct, or dormant, retain residual heat in the crust. This produces other geological phenomenon such as hot springs, mud pots, and geysers. Groundwater that has percolated through a hot zone through tubes and joints, will rise to the surface forming a pool. These hot springs may have water several thousand feet below the surface. The hot springs that are full of minerals keep churning mud and the bubbling gas and mud rise to the surface to form mud pots. Geysers are hot springs that spew hot water and steam, then stop. Cooler groundwater enters a hot spot and is heated. The escaping water and steam pass through the vent until that amount of water has been processed. The geyser stops until the next volume of cool water is heated and rise to the surface.

Old Faithful

The landforms that show horizontal layering, including the deepest valleys, are referred to as the plains and plateaus. The layers are filled with materials that have been eroded from mountains or by volcanic eruptions. In plains regions, the materials have not shifted from the depositing sites by uplift. Uplift that creates plateaus is located to the nearby mountains which may transform the plains into plateaus. The metamorphic rock that is leveled by erosion forms the plain and plateaus foundation.

Plains – are the horizontal layering without crustal disturbance. A plain is near the level of the last of the deposits that created it. If the plain is at the coast, the plain would be at sea level. It can be thousands of feet above sea level. Plains become dissected. The depth of the dissections is determined by the original altitude of the plain and the time from the deposition that formed it. High plains may have a little dissection. Their relief may be hundreds of feet. The Coastal plains form gradual valleys near sea level.

Alluvial plains – are created by stream deposition. They are smooth, sloping gently downstream. Deltas and floodplains are in this group classification. Alluvial plains could have started 50 million years ago. The deposition continued through the Pleistocene era. The wide distances between rivers preserve the alluvial flatness. Erosional remnants accumulate at major rivers.

Marine plains – are shaped by residual marine deposits above the water.

Glacial plains – are shaped by continental ice sheets. Ice will smooth terrains of low altitude and cover them with forms different from glacial activity.

Lake plains – are sediment covered lake bottoms that went dry which were caused by basin-filling or outlet-cutting, or the removal of glacier ice dams, or by the climatic change from humid to arid weather. Lake plains spread out over area glaciated in the Pleistocene era. When a river valley was dammed by ice to make a lake, the lake floor was exposed when the glacier melted. In some areas, when the lake dries up and the mountain streams continue to flow, the minerals remain through the evaporation phase. Salt from mountain streams deposit this mineral and when the lake begins to expand, the amount of salt turns what would normally be a freshwater lake into a salt lake.

Lava plains – are lands leveled by the filling of valleys from free-flowing lava or the pyro-clastics emissions. Upon close examination, lava tunnels, spatter cones, or pressure ridges may be evident.

Sand-dune plains – is created by the wind where vegetation is sparse and sand is abundant. These plains form on beaches and deserts as well as arid floodplains.

Loess plains – are similar to sand-dune plains. Loess plains are shaped by the wind blowing sand, silt, or dust from deserts. These materials can accumulate to a depth of 40 feet.

Coastal plains – are sediment-covered areas of the continental shelf that has recently emerged above sea level. They are cut by seaward running, parallel streams in valleys joined at a wide angle by tributaries. Some plains end on the landward side in a lowland. Plains often contain lakes or marshes in shallow basins created by crustal warping or wind erosion. A plain's landward edge is straight and indented at the valley mouths.

Sedimentary plateaus – The uplift of geosyncline deposits of medium thickness forms a sedimentary plateau. These plateaus have not experience deformation and removal of sediments. However, the sedimentary plateaus are dissected. The beds were lifted at the same time as the deformed mountain areas but remained horizontal. Sedimentary plateaus are part of the mountain building process as evidenced by broad warps, open folds, and monoclines as well as faults and marine beds above sea level.

Plateaus include the foreland type near the geosyncline margin and at the median type that is on a rigid block high in the geosyncline interior. After the uplift, a low part of a plateau is covered with sediments that formed on an overlapping plain. Very deep erosion of a plateau reveals the geosyncline basement which is the original igneous or metamorphic rocks on which the sediment was deposited.

Lava plateaus – Volcanic activity over thousands of years will fill a basin with lava to a depth of thousands of feet. Uplift will make this a lava plateau. Well-jointed rock and loose materials contain groundwater which emits springs from cliffs.

Deserts are terrains with less than 10 inches of rainfall a year. Deserts comprise one-third of the Earth's land. The desert climate ranges from hot as in northern Africa, to cold, as in the Arctic. The lack of rain is due to the remoteness from the ocean as in central Asia. Another reason is that the moisture in the air is removed by the cooling and precipitation that occurs as the moist air rises over mountainous areas. Other reasons include the lack of snow melting and the heating as it descends with the increase of pressure at the horse latitudes.

Desert land is shaped by streams despite the arid climate. Rare rains come with a sudden downpour producing brief torrents. The winds polish and sculpture rock as the produce sand dunes. Weathering with its dryness slowly reduces the rock to fragments which forms the deserts' loess deposits and sand dunes. Rock waste becomes part of the erosion process via streams. Water tables are very low. During the Pleistocene era lands that have now desert was once more humid.

Desert stream process – During the rains on rocky or hard-packed desert ground remain on the surface. The sheet of water sweep the slopes with weathered rock litter. This accumulates in gullies. The water may come in torrents as it rushes down toward the valleys. The torrents become mudflows or dissipate due to evaporation or seepage into the terrain. There are very view streams to feed springs. The water table is very deep.

Desert valleys – which includes small trenches called 'arroyas', are steep-sided because the streams are torrential and vigorously scrape the channel with debris. During long periods between rains, alluvium is exposed on valley floors. This allows the wind to form sand dunes and dust for dust storms.

In arid plains, the weaker rock is dissected by streams creating badland topography. Networks of V-shaped gullies are formed. The stronger rock forms the canyon with vertical walls.

Features created by wind erosion – Ripple marks are created by the eddies in a steady, gentle wind that moves sand grains forward on the ground. The stronger winds cause grains to jump one or two feet off the ground. Undercutting by abrasion is possible along the base of valley walls. Sand by a series of jumps, move up valley walls of a gentle slope. The dust from the sand abrasion is then washed by streams onto playas, then moved high and far from its source.

Pedestal rocks – look like mushrooms because of their wide topped pillars cut into the rock mass by stream erosion. These rocks are shaped by stream erosion or sandblasting. The narrow base is created by the intense abrasion near the ground.

Ventifacts – is Latin meaning 'made by the wind'. Ventifacts are loose stone shaped by the sandblasting. Facets are triangular in shape.

Desert pavement (Serir) – is closely packed pebble sized ventifacts that cover the desert bedrock. They are too heavy for the wind to carry.

Desert varnish – is highly polished or sandblasted rocks which have a coating of iron or manganese dioxide.

Etched rocks – are rocks with differently sandblasted surfaces. The softer minerals have eroded out of the rock. The hard rock materials remain. Etching usually honeycombs the rock shape.

Desert windows – are openings cut through the narrow, finlike ridges. It may appear to be a natural bridge but more likely it is eroded from a stream causing the tunneling effect.

Alcoves – are shallows cut in the base of a cliff by sandblasting or stream activity.

Blowouts – are made by the wind using the loose material. If a blowout reaches the water table, an oasis may appear. Oasis also occurs when the groundwater reaches the surface because of a fault.

Dunes cover only a small fraction of deserts. They build up where the terrain is flat, there is plenty of sand, and vegetation is sparse. Dunes form on deserts and on sandy beaches as well as on semiarid floodplains. Sometimes vegetation will stabilize the dune, but the winds wear down the dune and the sand migrates to a canyon, oases, up mesa sides, or burying a forest.

Barchans – are dunes of a crescent shape. The wind causes these dunes to form a slope of 30 to 35 degrees. The slip face is then hollowed out by eddies. Barchans form in swarms near the edge of the sand-plain where the sand is thinning.

Parabolic dunes – are like barchans but have longer horns pointing windward.

Transverse dunes – are long irregular bands at right angles to the prevailing wind.

Seifs – are long sharp-ridged dunes in the Arabian and Saharan deserts.

Whalebacks – are dunes with a broadly rounded back. These dunes are found in the Sahara. They have a flat top and they are the largest of all dunes. They can reach 150 feet high and 2 miles wide.

Major Deserts			
Name	Location	Length (miles)	(kilometers)
Sahara	North Africa	3,500,000	9,065,000
Gobi	Asia	500,000	1,295,000
Libyan	Africa	450,000	1,165,500
Rub al-Khali	Africa	250,000	647,500
Kalahari	Africa	225,000	582,800
Great Sandy	Australia	150,000	338,500
Great Victoria	Australia	150,000	338,500

Dust Storm

NOAA

Beach Monitoring

Where the sea meets the land, geological changes occur rapidly. It extends inland as far as tidal water travels. In some areas, the tidewater extends into estuaries as far as 150 miles. A 'shore' is an area where the water line migrates as the sea level changes with the tides and storms. It is called a shoreline when the area is narrow. Coasts are constantly influenced by tides and currents which shapes the shoreline.

Types of shorelines

Neutral shorelines – develop on coasts made by a deposition above sea level. Many indented shorelines are located on bird-foot deltas since this is where the stream deposits are more than wave action. Coasts of glacial deposits develop shorelines with shapes influenced by the deposits left.

Shorelines of emergence – are the coasts that consist of sediment covered areas of low relief that recently emerged from the sea. If the sediment was spread evenly by sea activity, then the shoreline would be straight. If the deposits were uneven, then the shoreline becomes irregular.

Shorelines of submergence – develop on dissected mountain or massif slopes that became submerged by a rise of sea level. The shoreline is located along river valley walls. Where the structural grain is oblique to the shoreline, there are estuaries, headlands, and islands.

Rocky beach

Wave Activity

Waves that are not generated by the wind are responsible for rapid changes in sea level.

Ocean tides – are produced by the Moon's gravitational attraction on Earth. A tide will crest every 12 hours. This allows the wind generated waves to work over vertical zones.

Storm surges – are the several irregular rising and falling of the sea level which is due to atmospheric pressure changes.

Tsunamis – are occasional large wave movements caused by sudden earth movement of the seafloor. The sea will retract and then suddenly surge with a high-level wave that is usually very destructive.

Ocean waves – are made at sea by the wind. As surface water is pushed forward, water rises from below to replace it. A circular motion is created. Every water particle moves forward at the surface, then down, back, up, and forward again. With each revolution, the wave moves forward in the direction of the wind. The waveform and size depend on the force of the wind and the distance it must travel. This produces what is called a 'swell'. Erosion by these waves is dependent on the back and forth movement of the wave.

Breakers – are waves that enter a shallow area and the circular motion of the wave is broken. The breaking wave curls and crashes. The breakers fall apart as the wind comes from the opposing direction. Spilling breakers crumble gradually and are made by steep, wind backed waves of a short length. Breaking waves have a turbulence which moves water and sediment toward the beach where the wave dissipates as sheets of sea water. Most of the water returns to the sea as backwash. Some of the sea water is absorbed by the sand. Oblique swash and backwash create beach drifting and may move sediment along the shoreline.

Waves affect the shore by impact, washing, wedging by forcing water into cavities and joints under pressure, suction due to the withdrawal of water, and corrosion which is rubbing or knocking of rock fragments against one another. The wave action is continuous and becomes destructive to the shore.

Waves breaking on a sea cliff will erode the lower portion of the cliff producing an overhang. These break off and become corrosive tools for further wave erosion. Eventually, the cliff retreats and a rock platform called a 'bench' takes form and widens beneath it. Large rock fragments are rounded and reduced to cobbles, gravel, pebbles, sand, and silt, thus creating a beach.

Waves breaking keep working the sediment where it transports it downwind.

Currents

Currents gather and transport sand and other consolidated material while eroding some areas and building up deposits in other areas.

Tidal currents – When the tide rises, water penetrates inland via channels. When the tide falls, water withdraws to the sea. The greater the rise and fall of the tidal water, the greater the velocity and erosive effect. Tidal currents shape coastal lowlands.

Rip currents – are fast-moving water flows that are caused by water from breakers and longshore currents returning to the sea. These rip currents travel rapidly, hundreds of feet to the sea. They contain churning sand.

Longshore currents – run parallel and close to shore. This is caused by waves impinging on the shore obliquely.

Wave-cut benches – are rock platforms that extend seaward from the base of cliffs. They are cut by sediment dragged over the bottom by waves and currents. The cutting is at the level of the storm waves.

Sea grottoes – are caverns shaped in limestone by groundwater and marine activity.

Wave-cut notches – are sharp indentations in cliffs made by waves near the waterline.

Sea caves – are cavities developed in weak rock in wave-cut notches.

Blowholes – are cave openings through which air is compressed by incoming waves and is then blown out through the opening in the cave roof as the wave enters the cave, or as water is withdrawn.

Sea arches – are the remains of headlands cut through by waves and currents.

Sea stacks – are pillars and columns left by the erosion of wave-cut benches. Some are remains of collapsed arches.

Beaches

Beaches involve the shifting of sediment along the shore. The sloping deposits of sediment are shaped by waves and the current actions of currents and storms. On the rocky coast, beaches form at indentations where the sediments are moved along the shore and become trapped.

Beach sediment comes from various conditions such as marine erosion of cliffs, or volcanic materials, or glacial drift, or reworked stream deposits. Sand may be composed of ground coral and shells, or quartz, feldspar, magnetite, or garnet. The various rocks found at the beach are in the form of pebbles or cobbles. The landward direction of the beach will consist of a cliff or a gradual slope beyond the reach of storm waves.

Swash marks – are thin, low ridges of sand, seaweed, or other marine debris left by the highest reach of the swash.

Ripple marks – are wavy, parallel ridges of sand or silt formed by the waves or currents. They are exposed at low tide.

Rill marks – are small drainage channels cut by water from waves or seepage at low tide.

Beach cusps – are low mounds pointed on the seaward side between the drainage channels near the water level. They are shaped by waves and currents in the tidal area.

Beach ridges – are ridges of gravel or cobbles built up by storm waves on the landward side of the backshore.

Sand barriers – are masses of sand carried by waves and longshore currents. They are deposited above or below the tide line where sand is thick.

Barrier islands – are broad, long-lasting sand barriers built well above the high tide by marine action and wind.

Spits – are sand barriers connected to the shore at one end.

Lagoons – are areas of quiet water between a barrier island and the mainland. Lagoons may eventually fill with sediments.

Tombolos – are sandbars that connect islands to the mainland.

Cuspate Forelands – are wedge-shaped deposits built out from the shore at points where the coast movement of sediment, or eddy points, and the current, turn seaward.

Organic reefs – are ridges built up from the sea bottom by the deposition of hard parts of corals or similar organisms.

** For more information on beaches, beach evaluation, or oceanography badge requirements, obtain a copy of **BEACHES**, by Jack Fleming **

Lakeshores – Lakes are usually short-lived because they fill with sediment and they empty out through outlet streams and evaporation. Because of the short time of existence, the lake doesn't have enough time to shape the shoreline. On a large lake, the shoreline may be sculptured in weak rock formations and wave-cut cliffs and terraces. The shores of kettle lakes are smooth and round by wave and current activity. The beaches of very large lakes will show the same features of typical sea beaches.

Shoreline evolution – Prolonged marine activity without earth movements or glaciation will permit the shore to transcend through a natural progression of development. The sea tends to shape the shoreline into broad curves. Rocky shorelines first develop into cliff headlands and the eroded material is deposited as spits. Continued erosion broadens the beach. Shorelines that are smooth continue to remain smooth. Low lying emergent shores produce barrier islands in the shallow water. Eventually, they move back toward the mainland and merge with it.

Major Lakes			
Name	Location	Average Area Sq. Miles/ Kilometers	Average Depth Feet Meters
Caspian Sea	Asia	143,630 372,002	3,264 995
Lake Superior	North America	31,700 82,103	1,330 405
Lake Victoria	Africa	26,828 69,485	270 82
Lake Huron	North America	23,000 59,570	750 229
Lake Michigan	North America	22,300 57,757	923 281
Aral Sea	Asia	15,500 40,145	177 54
Tanganyika	Africa	12,700	4,708

Name	Continent	Area		Depth	
		32,893		1,435	
Baykal	Asia	12,162		5,315	
		31,500		1,620	
Great Bear	North America	12,096		1,299	
		31,329		396	
Nyasa	Africa	11,150		2,280	
		28,879		695	
Great Slave	North America	11,031		2,015	
		28,570		614	
Lake Erie	North America	9,910		210	64
		25,667			
Winnipeg	North America	9,417		60	18
		24,390			
Ontario	North America	7,550		802	244
		19,555			
Balkhash	Asia	7,115		85	26
		18,428			
Ladoga	Europe	6,835		738	225
		17,703			
Chad	Africa	6,300		24	7
		16,317			
Maracaibo	South America	5,217		115	35
		13,512			
Onega	Europe	3,710		328	100
		9,609			
Eyre	Australia	3,600		52	16
		9,324			
Volta	Africa	3,276		Unknown	
		8,485			
Titicaca	South America	3,200		922	281
		8,288			
Nicaragua	North America	3,100		230	70
		8,029			
Athabasca	North America	3,064		407	124
		7,936			

Major Rivers			
Name	Location	Length (Miles) (Kilometers)	
Nile	Africa	4,160	6,693
Amazon	South America	4,000	6,436
Yangtze	Asia	3,964	6,378
Mississippi (Mo)	North America	3,740	6,017
Huang (Yellow)	Asia	3,395	5,463
Ob-Irtysh	Asia	3,362	5,409
Amur	Asia	2,744	4,415
Lena	Asia	2,734	4,399
Congo	Africa	2,718	4,373
Mackenzie	North America	2,635	4,240
Mekong	Asia	2,600	4,183
Niger	Africa	2,590	4,167
Yenisey	Asia	2,543	4,092
Parana	South America	2,485	3,998
Mississippi	North America	2,340	3,765
Missouri	North America	2,315	3,725
Murray-Darling	Australia	2,310	3,717
Volga	Asia	2,290	3,685
Purus	South America	2,100	3,379
Madeira	South America	2,013	3,239
Sao Francisco	South America	1,988	3,199
Yukon	North America	1,979	3,184
Rio Grande	North America	1,900	3,057

Geologist coring a lake

Geology is a broad field with many career possibilities. A hydrogeologist searches for water underground. He will work with planning water use and with construction workers building foundations for buildings. Geological engineers' advice where to build major structures such as bridges and dams. They are in many various construction sites from build piers at a beach to build tunnels through mountains. Geologists are also involved with searching and the production of natural resources. Oil, gas, and coal exploration and product are a major concern throughout the world. Metals such as iron, copper, and gold are always in demand as well as other minerals.

General requirements to become a professional geologist are that the individual earns a four-year college degree in geology, geophysics, environmental geology, or geological engineering. In certain fields, a Master's Degree is required. If one chooses to teach at a college, then a doctoral degree may be required. The college student studying the field of geology will include other science courses such as chemistry, physics, and mathematics. The knowledge of computers is required now in almost every field of study.

Careers and Fields of Study in Geology

Economic Geology – is a very important branch of geology. This field of geology deals with the economic minerals being used by industry to fill various needs by extracting the minerals economically while obtaining a profit. The economic geology locate the resources and manage the extraction such as coal and oil as well as iron, copper, uranium, and gold.

Mining Geology – is the science of extracting resources from the Earth. These geologist extract resources such as gemstones, metals, asbestos, mica, clay, quartz, zeolites, phosphates, perlite, chlorine, helium, sulfur, and silica.

Petroleum Geology – are geologist employed to study locations for oil and natural gas. Because these resources are located in sedimentary basins, the study includes sedimentary and tectonic formations including the present day formations of the rock formations.

Engineering Geology – is the study of geologic principles to engineering for the application affecting design, construction, location, maintenance, and operation of engineering projects.

Civil engineering is the study of geological principles as applied to engineering of the buildings and structures. This area of study includes bridges, buildings and foundations, tunnels, and more. The engineering geologist assures construction of major projects are constructed safely and built to last.

Hydrogeology – is the study of groundwater. The geologist analyses specific locations and searches for groundwater which can supply uncontaminated fresh water. The geologist maintains the water supply to monitor the spread of any contamination in groundwater wells. These geologist use various methods to obtain data. They use stratigraphy, core samples, ice cores, sediment cores, and boreholes.

Natural Hazard Geology – are geologist and geophysicists that study natural hazards in order to maintain safe building codes and warning systems pertaining to geological hazards. These hazards include: earthquakes, avalanches, sinkholes, rock-fall, volcanoes, mudslides, floods, tsunamis, river channel and migration, landslides and debris flow, and more.

Today with the growing industries worldwide, the need for geologist is increasing. Geologist are an important part of numerous major projects.

Resources:

American Association of Petroleum Geologist, 1444 South Boulder, Tulsa, OK 74119

800-364-2274   website: http://www.aapg.org

American Geosciences Institute, 4220 King St. Alexandria, VA 22302-1502

703-379-2480   website: http://wwwagiweb.org

American Petroleum Institute, 1220 I. Street NW, Washington, D.C. 20005-4070

202-682-8000   website: http://www.api.org

The Geological Society of America, P.O. Box 9140, Boulder, CO 80301-9140

303-357-1000   website: http://geosociety.org

U.S. Geological Survey, website: http://www.usgs.gov

Paleontological Research Institute, 1259 Trumansburg Road, Ithaca, NY 14850

607-273-6623 website: http://www.priweb.org

1. What is the definition of geology? Why is the study of present geology important to understanding the past geology? How do geologist learn about rock formations?
2. Name three resources that can be mined from the Earth for commercial use. Explain how each of the three products are discovered and processed.
3. Study a geologic map of your area and identify the different rock types and estimated ages of the rocks. How can you tell if the rocks are faulted, horizontal, or folded?
4. Select ONE of the following:
    a. List three career opportunities in geology. Select one and explain the education, training, and experience required for that position. Why would you be interested in that particular profession?
    b. If possible, visit a land-use planner, civil engineer, or geologist. Discuss what projects they are working on and what type of equipment is used. Ask about their specific tasks with maps and reports created for this project.
5. Select ONE of the grouped requirements, (A, B, C, or D).
    a. Surface and Sedimentary Processes.
        i.   In an experiment, demonstrate how sediments settle from suspension in the water. Explain why this is important.
        ii. With a topographical map, locate a stream and identify if the stream is dendritic, meandering, trellis, or straight. Showing the different elevations, explain which streams flow faster and why.
        iii. Using a stream diagrams, show each feature: cut bank, fill bank, medial channel

bars, point bar, and lake-delta. Explain sediment grain size at each feature.

    iv. Conduct an experiment where sedimentary material that is carried by water is too small to see without magnification.

    v. Visit a nearby stream. Even if the stream has dried, record clues of the direction of the stream in a notebook. Sketch as much detail as possible.

b. Energy Resources

    i. List the five Earth resources that are used to generate electrical power.

    ii. Explain the three components for the occurrence of oil and gas in the ground. The source of rock, trap, and reservoir rock are the areas to identify.

    iii. Explain each of the items used in subsurface exploration to locate oil or gas: electric well logs, reflection seismic, offshore platform, geologic map, stratigraphic correlation, subsurface isopach map, subsurface structure map, and core samples as well as cutting samples.

    iv. Create a 20 point subsurface map and explain how geology maps are used to find oil, gas, or coal resources.

    v. Select ONE of these two activities:

        1. If possible, arrange to visit an operating drilling rig. Ask to see the geologist and request to see a core sample. Ask what the geologist does on the site.

        2. Create a display showing how oil and gas or coal is found, extracted, and processed. Research your information using books, maps,

articles, or internet information. Be prepared to give a five-minute explanation of your display.

c. Material Resources.

    i. Define Rock. Explain the three classes of rocks, their characteristics and origin.

    ii. Define mineral. Explain the origin of minerals. What is their chemical composition and how do you identify the mineral. Include there hardness, streak, color, specific gravity (density), luster, cleavage, and crystal form.

    iii. Select ONE of the following:

        1. Identify 15 different rocks and minerals. Label each specimen. If it is a rock, name its class. If it is a mineral, label its physical properties.

        2. Collect 10 different rocks or minerals. Record each finding in a notebook. Identify each specimen including its class and origin. Show its chemical composition. List its physical properties.

        3. List three of the most common road building materials used in your area. Explain how each material is produced and used in road building.

    iv. Do ONE of the following:

        1. If possible, visit an active quarry, mining site, or gravel-sand pit. Explain how these resources are used.

        2. Discuss two examples of rocks and two examples of minerals. How are these material mined and how are they used.

3. If possible, visit the office of a civil engineer and learn how geology is used in construction.

D.   Earth History.

1.   Explain the process of burial and fossilization. Discuss extinction.

2. Create a chart showing suggested geological eras and periods. Determine which period the rocks in your area might have formed.

3. Explain how fossils provide information about ancient life, the climate, the environment, and geography. Explain how food was obtain in the following eras: benthonic, pelagic, littoral, lacustrine, open marine, brackish, fluvial, eolian, and protected reef.

4. Identify 15 different fossil plants or animals. Record the information in a notebook. List where the items were found. Explain how each survived.

5. Do ONE of the following:
   a. Visit a geology department of a local university or visit a science museum. If possible, arrange to have a curator guide you

through the fossil preservation process.

b. Find a structure in your area that was built with fossiliferous rocks. Make note of the fossil evidence that you find.

c. Visit an area with a fossil outcrop. What type of rock contains the fossils? What type of fossils did you find?

d. Create a display on the common fossils found in your state. Label the date on an image of the fossil. Research your selected fossil. Use maps, books, periodicals, and the internet to prepare your presentation. Discuss your presentation with a group.

"Knowledge is the light of life"

...share the light!

Check out other books by Jack Fleming

BEACHES, by Jack Fleming - contains beach evaluation, Oceanography badge requirements, experiment, activities, and much more.

WEATHER, by Jack Fleming — contains loads of information regarding weather prediction. Meteorology badge requirements, experiments, fun activities, and more.

THE ALIEN PLAN, by Jack Fleming. His first book of fiction. The worst fear of every child is the loss of a parent. Kevin Cahill learns the value of family and friends. When all hope is lost, Kevin and his friends find a way. Eventually, Kevin has to make a difficult decision whether to leave home or stay.

www.ingramcontent.com/pod-product-compliance
Lightning Source LLC
Chambersburg PA
CBHW071310220526
45468CB00001B/313